INTEGRATING ENGINEERING + SCIENCE IN YOUR CLASSROOM

Edited by
Eric Brunsell

National Science Teachers Association

Arlington, Virginia

National Science Teachers Association

Claire Reinburg, Director
Jennifer Horak, Managing Editor
Andrew Cooke, Senior Editor
Wendy Rubin, Associate Editor
Agnes Bannigan, Associate Editor
Amy America, Book Acquisitions Coordinator

ART AND DESIGN
Will Thomas Jr., Director
Joe Butera, Senior Graphic Designer, cover and interior design
Cover photograph by Joan Cantó

PRINTING AND PRODUCTION
Catherine Lorrain, Director
Jack Parker, Electronic Prepress Technician

NATIONAL SCIENCE TEACHERS ASSOCIATION
Gerald F. Wheeler, Interim Executive Director
David Beacom, Publisher
1840 Wilson Blvd., Arlington, VA 22201
www.nsta.org/store
For customer service inquiries, please call 800-277-5300.

NSTA is committed to publishing material that promotes the best in inquiry-based science education. However, conditions of actual use may vary, and the safety procedures and practices described in this book are intended to serve only as a guide. Additional precautionary measures may be required. NSTA and the authors do not warrant or represent that the procedures and practices in this book meet any safety code or standard of federal, state, or local regulations. NSTA and the authors disclaim any liability for personal injury or damage to property arising out of or relating to the use of this book, including any of the recommendations, instructions, or materials contained therein.

Library of Congress Cataloging-in-Publication Data
Integrating engineering and science in your classroom / Eric Brunsell, editor.
    pages cm
  ISBN 978-1-936959-41-9
  1. Science--Study and teaching (Elementary) 2. Engineering--Study and teaching (Elementary) 3. Science--Study and teaching (Secondary) 4. Engineering--Study and teaching (Secondary) I. Brunsell, Eric.
  LB1585.I495 2012
  372.35'044--dc23
                        2012026196
  ISBN 978-1-936959-63-1 (ebook)

Contents

Contents

Part Two
Content Area Activities

LIFE AND ENVIRONMENTAL SCIENCE

EARTH SCIENCE

PHYSICAL SCIENCE

Contents

Part Three
After-School Programs

Introduction

I still remember my very first day as a teacher. A few days earlier, my principal had given me this advice: "Whatever you do, do not start with an overview of your course. Do something active and set the tone. The overview can wait for day two or three." Of course, I had planned on introducing the course, so I had to quickly make some changes. I decided to have students do a simple tower building challenge. As I nervously awaited the arrival of my first period physical science students, I wondered how they would react. To my relief, they jumped right in and remained engaged throughout the entire class—building, testing, and revising prototypes of paper towers. Since that day, I have continued to use design challenges as a way to engage students and teachers in classes and professional development. I have also been involved in other after-school engineering clubs and competitions. Almost universally, these activities engage and excite students of any age. More importantly, when done well, these activities reinforce important skills and science content. The recent surge in interest for engineering education, including engineering's inclusion in the *Framework for K–12 Science Education* and *Next Generation Science Standards*, presents an exciting opportunity for teachers of science. This book includes a variety of excellent articles from NSTA's *Science and Children*, *Science Scope*, and *The Science Teacher* to help you integrate authentic and meaningful engineering activities into your teaching.

Engineering in the NAS Framework for K–12 Science Education

The National Research Council's Framework for K–12 Education states, "any science education that focuses predominately on the detailed products of scientific labor—the facts of science—without developing an understanding of how those facts were established or that ignores the many important applications of science in the world misrepresents science and marginalizes the importance of engineering" (2012). The Framework continues by identifying two implications that this statement has for science education standards:

- Students should learn how scientific knowledge is acquired and how scientific explanations are developed.

- Students should learn how science is used, in particular through the engineering design process, and they should come to appreciate the distinctions and relationships between engineering, technology, and applications of science.

The Framework includes two engineering "core ideas" for K–12 science education. ETS 1: How do engineers solve problems? and ETS 2: How are engineering, technology, science, and society interconnected? The first core idea focuses on students understanding an engineering design process that includes defining an engineering problem, developing potential solutions, and optimizing the design solution. The second core idea focuses on helping students understand the interdependence of science, engineering, and technology, and their influence on society and the natural world. The Framework also describes a series of eight science and engineering practices, summarized in Table 1.

The inclusion of engineering concepts and practices in the Framework is not intended to add more to the plate of teachers with an already overburdened science curriculum. Additionally, it is not intended to replace state or district engineering standards nor co-opt stand-alone engineering courses and programs. Instead, these core ideas and practices are meant to help teachers introduce the interdependence of science, technology, and engineering and to harness the power of design activities to support authentic learning of science concepts.

Including engineering activities in your science curriculum can reinforce science concepts while providing experiences for your students that illustrate a wide range of STEM skills, issues, and opportunities. For example, in Yocom De Romero, Slater and DeCristofano's, "Design Challenges Are ELL-ementary," students learn about properties of Earth materials as they design a wall. High school students explore gene therapy technology and issues in Lockhart and Le Doux's article, "A Partnership for Problem-Based Learning." In Thompson's "The Science of Star Wars: Integrating technology and the Benchmarks for Science Literacy," middle school students apply their understanding of electric circuits as they design and construct model light sabers.

In their article, "Engineering for All," Lottero-Perdue, Lovelidge, and Bowling describe the

---

Engineering Core Ideas in *A Framework for K–12 Science Education*

ETS 1: How do engineers solve problems?

- Defining and delimiting an engineering problem
- Developing possible solutions
- Optimizing the design solution

ETS 2: How are engineering, technology, science, and society interconnected?

- Interdependence of science engineering, and technology
- Influence of engineering, technology, and science on society and the natural world

www.nap.edu/catalog.php?record_id=13165

---

implementation of an engineering design challenge in an inclusive environment. They end with two warnings, "(1) brace yourself for the excitement that students have as they engage in the engineering design process; and (2) be prepared for all students to succeed and for some who normally struggle to shine." The hands-on nature of engineering design activities will engage your students, foster higher-order thinking, and deepen their understanding of how science, technology, engineering, and mathematics (STEM) influences the world around them. By exposing students to authentic engineering activities, you can help students uncover the profession that makes the world work.

TABLE 1.

The Framework's eight science and engineering practices

Practice	Description
Asking Questions and Defining Problems	Engineering challenges usually start with a need, want, or problem. Engineers ask questions to help define the problem by identifying constraints and identifying criteria for success.
Developing and Using Models	Models can be physical, conceptual, or mathematical. Engineers use models to identify and test solutions.
Planning and Carrying Out Investigations	Engineers use investigations to test and refine their solutions. Engineers need to be able to identify variables and develop investigations to test the reliability or capability of their solutions.
Analyzing and Interpreting Data	Engineers need to be able to analyze and interpret data by using graphs and other representations of data to determine the suitability of their solutions.
Using Mathematics and Computational Thinking	Engineers use mathematics to develop models and test solutions. Engineers also use computers to assist with data analysis and simulations.
Constructing Explanations and Designing Solutions	Engineers use the engineering design process to develop solutions for problems. Engineers must balance many different factors (e.g., cost, esthetics, materials, safety) as they develop solutions. Additionally, engineers must make judgments about which solution might be the most fruitful depending on specific criteria.
Engaging in Argument From Evidence	Argumentation is the use of reasoning to create a claim supported by evidence. Engineers must be able to craft an argument to explain and defend their design decisions. Engineers should also be able to critique arguments created by others.
Obtaining, Evaluating, and Communicating Information	Engineers must be able to read and comprehend technical information from a variety of sources, including text. Additionally, engineers must be able to effectively communicate ideas and collaborate with others.

Organization

This book begins with an initial essay on engineering design as a problem-solving approach. Next, the book is divided into three major parts. Part one, Engineering Design, illustrates the engineering design process in elementary, middle, and high school courses. Part two, Content Area Activities, provides ideas for supporting disciplinary content in life and environmental science, Earth science, and physical science by using engineering concepts and processes. Each subsection includes a diversity of articles at each grade level. Part three, After-School Programs, examines model after-school programs that can engage students in science and engineering activities.

National Research Council (NRC). 2012. *A framework for K–12 science education: Practices, crosscutting concepts, and core ideas.* Washington, DC: The National Academies Press. Available online at *www.nap.edu/catalog. php?record_id=13165*

PART ONE
Engineering Design

CHAPTER 1

The Engineering Design Process

By Eric Brunsell ///

Students need more than one chance to be successful at a task. So many times they are left thinking, "next time I would have…." The design process allows students to have that next time.

—Jonathon W. Gerlach
"Elementary Design Challenges"

Science and engineering are complementary, but they are not the same. In general, the purpose of science inquiry is to use evidence to explain the natural and designed world. The purpose of engineering is to solve specific problems related to needs or wants. Many different characterizations of the design process can be found in the articles in this book. However, they all share similar aspects that involve defining the problem, generating multiple possible solutions; analyzing these solutions; testing, evaluating, and refining solutions; and communicating ideas. Engineering design is messy—it is not a linear process. Additionally, engineering design is not a "one shot deal." Failure is a constant companion in the design process. By testing ideas to failure, we learn how to improve our ideas. In the classroom, testing of a solution should not be the end point. Instead, like in Gerlach's quote, students should be given opportunities to refine their solutions as many times as possible within the constraints of your curriculum.

Define the Problem

In Sumrall and Mott's *Science Scope* article, "Building Models to Better Understand the Importance of Cost Versus Safety in Engineering," students are asked to build a tower for a football field. Identifying parameters for a successful tower and identifying the need to balance safety and cost further define this problem. In this project, the balance between safety and cost helps students understand "trade-offs" inherent in the decision-making process.

Often, engineering projects are started with a design brief. A design brief includes the goal of the project, expectations, and limitations. The design brief can be introduced by the teacher (as in Sumrall and Mott's project) or collaboratively created by students and the teacher. An example design brief can be found in Sterling's "Science and engineering" article.

As part of this stage of the engineering design process, it is common for students to conduct research about the problem and what solutions have been used in the past.

Develop Possible Solutions

In any complex problem, there is no obvious right answer. In fact, there are often many solutions. At this stage of the design process, students should brainstorm as many possible ways to solve the problem. Good brainstorming involves rapidly generating ideas without passing judgment. If a problem is particularly challenging, it is often valuable to break the problem into smaller pieces and brainstorm solutions to each piece. For example, if the challenge is to create a robot that can move widgets from one assembly line to another, you might brainstorm solutions for moving the robot separate from solutions for grabbing widgets.

Brainstorming should be a social process. It may help to start the process if students identify a few possible solutions on their own, but most of the brainstorming time should be spent in small groups. To deemphasize competition, it is also beneficial to continue brainstorming as an entire class. As you facilitate this process, help students keep their focus on generating ideas, not evaluating the likelihood that the idea is suitable.

Analyze Solutions

Once a large number of possible solutions have been identified, students can begin analyzing solutions. In this process, students make judgment calls on how each solution meets the goals and constraints (specifications) in the design brief. Students should be encouraged to use a systematic approach for this analysis. For example, a simple matrix can be used to determine if a solution meets the required specifications. From this analysis, students select one (or a few) solution to prototype and test.

In Sakakeeney's article, "Repairing Femoral Fractures," students are challenged to find ways to mend a broken leg. Before getting their hands on a model leg, students are expected to create two possible design solutions. The design solution includes a description of how the solution works, how it matches the problem definition, and clear drawings. In small groups, students evaluate each of these design solutions by identifying pros and cons and then develop the solution that they plan on testing.

Optimize Solutions

The next step in the engineering design process is for students to test their solutions. Students should collect data on the performance of their solution and identify opportunities for improvement. In some cases, these tests can take the form of a "fair test," where single aspects of a design can be changed to determine how it impacts performance. At this stage, it is critical to reinforce that engineering design is an iterative process. Whenever possible, testing should lead students to refine their design. Students should be given the opportunity to retest those modifications as they improve their designs. This is time-consuming, but limiting testing to one attempt provides a distorted view of the engineering design process. Gerlach's "Elementary Design Challenges" illustrates this process of continual improvement as students work as a class to perfect airplane designs.

Solutions	Specifications		
	Criteria 1	Criteria 2	Criteria 3
Solution 1	X	+	+
Solution 2	–	X	–

Communication

Throughout the engineering design process, students should be presented with opportunities to communicate their results to their peers and, when possible, to a larger audience. During each stage, students can share their thinking. For example, in the "Define the Problem" stage, students can share their understanding of the scenario, constraints, and indicators of success for solutions. During the "Test, Evaluate, and Refine" stage, students should share the results of their test, decisions that they made related to refining their solution, and evidence that supports those decisions. Finally, students should be given the opportunity to develop a presentation of their final solution, evidence of success, and their pathway to that solution.

CHAPTER 2
Science and Engineering
Two Models of Laboratory Investigation

By Jennifer Harkema, James Jadrich, and Crystal Bruxvoort ///

The following scenarios describe lab activities commonly performed in high school science classes:

- Scenario A: Students studying gravity determine the value of the gravitational constant (g) by dropping balls from various heights and timing how long it takes them to hit the ground. Lab grades are based on how closely results compare to the expected value (9.8 m/s^2).

- Scenario B: Students build a model rocket fueled by water and effervescent tablets. Points are awarded based on the height reached by the rocket. Extra credit is awarded to the student whose rocket reaches the highest height.

- Scenario C: Students design an experiment showing that light intensity affects the rate of photosynthesis in Elodea.

- Scenario D: Students determine the density of various objects.

Laboratory activities have an important place in the science classroom. They introduce students to science concepts and at the same time, teach them about scientific processes and the nature of science (McComas 2005). While the value of these activities is clear, teachers

FIGURE 2.1.

Science model example

The following lesson was inspired by Scenario A (p. 7) at the beginning of this article. Instead of calculating the acceleration due to gravity, students determine what variables affect how fast an object falls. The teacher introduces this activity with a demonstration in which he or she drops a weighted film canister and a folder from about shoulder height. Students observe the film canister fall at a faster rate. After some discussion, they are presented two competing hypotheses:

1. Size (and shape) determine how fast an object falls.

2. Weight determines how fast an object falls.

Student groups are asked to test which, if either, of these hypotheses is correct. They must do this by designing fair tests (i.e., control variables) using film canisters, small washers, masking tape, and manila folders.

Some students may try to determine if weight matters by using two film canisters of the same size that contain different amounts of weight. These students conclude that

weight does not affect acceleration, as both canisters hit the ground at the same time. Other students will test the effect of weight by using identical folders, made unequal in weight by taping washers to one folder. In this case, they conclude that weight does have an effect. Some students may test for size and shape by simultaneously dropping a crumpled folder and a flat folder, leading to the conclusion that size and shape matter.

A large group discussion brings out the full complexity of the situation. Students cannot simply state that weight or size alone determines how fast an object falls because both variables, in certain cases, have an effect on the result. However, students can use their testing (along with the teacher's guidance) to conclude correctly that when objects are small in size—compared to their mass—weight does not matter. The lesson should not conclude without discussing important aspects of the scientific process, including fair testing, the tentative nature of scientific knowledge, and the use of disconfirming evidence to substantiate conclusions.

---

should be wary of incorporating too many lab activities like those described above. Although lab experiences like these are beneficial to students, they have the capacity to reinforce a false notion of scientific experimentation—namely that its purpose is to achieve a prescribed outcome. In this article, we focus on this misconception and offer suggestions for

how teachers can better portray the true nature of scientific experimentation to students.

Types of Experimentation: Science Versus Engineering

Understanding the nature of scientific experimentation is important in today's science classroom (AAAS 1990; AAAS 1993; NRC 1996). Scientific literacy—the goal of science education—requires not only a broad understanding of content, but also of the methods, purposes, and even limitations of scientific investigations.

As science teachers, we know that students often have preconceptions that are inconsistent with accepted scientific views and complicate science learning (Driver, Guesne, and Tiberghien 2002; Gomez-Zwiep 2008). Less well-known is that students' preconceptions also influence the way in which they perceive laboratory activities. As a result, some develop a false notion about the nature of scientific experimentation.

Scientific experimentation, rightly understood, is the examination of cause-and-effect relationships, with the goal of finding and understanding causal mechanisms in nature. This type of experimentation is referred to as the "science model" (Schauble, Klopfer, and Raghavan 1991). These types of experiments ask questions such as: "What affects reaction rates?" and "Is free-fall acceleration independent of mass?" Figure 2.1 presents an activity that transforms Scenario A (p. 7) into a science model investigation.

A second experimental approach is characterized by the manipulation of variables to produce a desired outcome. This type of experimentation is referred to as the "engineering model" (Schauble, Klopfer, and Raghavan 1991) and reflects, to a large degree, the inherent concern of engineers. Scenario

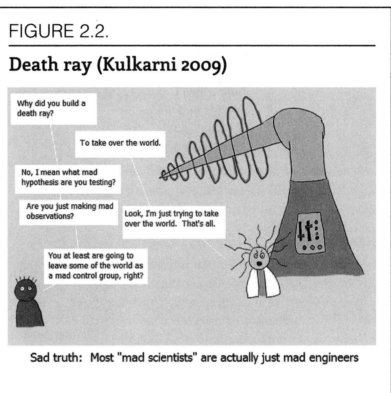

FIGURE 2.2.

Death ray (Kulkarni 2009)

Why did you build a death ray?

To take over the world.

No, I mean what mad hypothesis are you testing?

Are you just making mad observations?

Look, I'm just trying to take over the world. That's all.

You at least are going to leave some of the world as a mad control group, right?

Sad truth: Most "mad scientists" are actually just mad engineers

Popular media's portrayal of scientists and engineers at work can unintentionally reinforce misconceptions of both the science and engineering models of experimentation.

B (p. 7), in which students are asked to build a rocket that launches as high as possible, is one example.

Of course, practicing scientists and engineers often employ both the science model and the engineering model when conducting research. Dewey (1913) suggests that both approaches to experimentation are important—one in a practical sense for the purpose of achieving a desired effect (engineering model), and the other for the purpose of achieving scientific understanding (science model). Both models have an important place in the science classroom.

However, research has shown that students do not easily distinguish between the science model and the engineering model of experimentation (Kuhn and Phelps 1982; Schauble 1990; Schauble et al. 1991; Tschirgi 1980). Popular media portrayals of scientists and engineers at work can even unintentionally lead students to conflate these two models of experimentation (Figure 2.2, p. 9).

In most classroom labs, students are often predisposed to employ only the engineering model—when they should be using the scientific model (Schauble et al. 1991). For example, Schauble et al. (1991) report that students asked to investigate the effects of a car's design on speed became wrongly preoccupied with constructing fast cars (i.e., the engineering model). In this activity, the intention was for students to use the science model to investigate the effect of car design on speed; instead, they interpreted the purpose of the activity as optimizing a desired outcome—that is, designing fast cars.

Consequently, when teachers assign numerous lab activities that anticipate a prescribed outcome (e.g., show that $g = 9.8$ m/s^2 or demonstrate that light intensity affects photosynthesis), they reinforce the misconception that scientific experimentation is mainly about achieving a specific outcome—and not about finding relationships among variables. The challenge of achieving a specific outcome encourages students to use a sort of trial-and-error approach in the laboratory. Instead of thoughtfully applying concepts and designing experiments to explore cause-and-effect relationships, many students work toward attaining the prescribed result—and bypass the opportunity to develop important conceptual understanding.

Implications for Science Teachers

Students should understand and be proficient in using both the science and the engineering models of experimentation. Since students often are predisposed to using the engineering model, teachers

must intentionally design activities that better reflect the nature of scientific experimentation. We recommend considering the following guidelines when creating and implementing science labs:

- Caution against using verification labs: Labs in which students are asked to confirm an expected outcome portray science in ways that are at odds with how scientists investigate problems (McComas 2005; Lunetta, Hofstein, and Clough 2007). These types of activities encourage the misconception that the engineering model is the only model of experimentation. Therefore, the use of verification labs should be carefully balanced with the use of labs that employ the science model.

- Explore and apply: Instructional design should involve labs in which students first explore a concept by studying the relationships between causes and effects (Marek, Maier, and McCann 2008). Once students have developed an understanding of how important variables affect an experimental situation, they can be challenged to use the engineering model and apply their newly formed conceptual understanding to generate a product or maximize an output. In this manner, the science model is employed early on in the exploration phase of the lesson, and the engineering model is used in a subsequent phase of the lesson as an application of student understanding.

For example, consider Scenario B, in which students design rockets. This activity could be revised so that students first determine whether the amount of water, number of effervescent tablets, or fin size affects rocket height. This encourages students to focus on controlling variables to determine whether or not rocket height depends on any of these three variables. After students have distinguished the important variables, they can apply their knowledge by designing a rocket that will travel to the highest height.

- Careful design of research questions: Word choice can influence students' perceptions of the nature of science (McComas 2005). As an example, consider Scenario C (p. 7), in which biology students are challenged to show that light intensity affects the rate of photosynthesis. The wording of this question encourages students to use the engineering model. A slight rephrasing of the research question from "Design an experiment to show that light intensity affects the rate of photosynthesis in Elodea" to "What effect (if any) does light intensity have on the rate of photosynthesis in Elodea?" explicitly invites students to explore the possible effect of a variable.

- Careful use of competition in the classroom: Competitions, such as those described in Scenario B, can motivate students to really engage in science class (Fennema and Peterson 1987). However, activities involving competition in the classroom typically reinforce the engineering model. Such challenges should not be overused, but rather balanced carefully with activities using the science model.

- Test competing hypotheses: Students are taught to look for supporting evidence to justify their ideas (Nickerson 1998). However, if students are asked solely to present or find evidence for an idea, but are very rarely asked to construct evidence against an idea, they are led to believe that science is only concerned with confirmation—a form of the engineering model.

If students are asked to determine the value for the gravitational constant (g), (e.g., Scenario A, p. 7) or show that light intensity affects the rate of photosynthesis (e.g., Scenario C, p. 7), their task is mainly to seek confirming evidence. In science, however, theory making involves seeking confirming and disconfirming evidence. A balanced laboratory curriculum tasks students with testing competing hypotheses and generating evidence both for and against proposed scientific ideas.

For example, instead of asking students to calculate the acceleration due to gravity (Scenario A), students can be challenged to test the determining factor that affects how fast an object falls: the object's weight or its size and shape (Figure 2.1, p. 8). Similarly, instead of asking students to determine the density of various objects (Scenario D, p. 7), students can be challenged to test whether weight alone, size alone, or shape alone determines its density. Instead of doing straightforward exercises seeking confirmation of known results, students are challenged to engage in nontrivial exercises that better reflect the true nature and process of science.

Conclusion

Teachers want to teach correct scientific understanding and incorporate lab activities effectively and efficiently. As a result, they may feel compelled to overuse lab activities that favor the engineering model of experimentation. However, "accurate portrayal of the nature of the scientific endeavor stands at the core of all high-quality science teaching" (McComas 2005, p. 25).

Our goal as teachers is to design labs that reflect the nature of scientific experimentation and teach fundamental concepts, while simultaneously challenging students to use and understand both the scientific and engineering models of experimentation.

Acknowledgment

The authors wish to acknowledge support provided by National Science Foundation Grant No. DUE-0639694 and Calvin College.

References

American Association for the Advancement of Science (AAAS). 1990. *Science for all Americans*. New York: Oxford University Press.

American Association for the Advancement of Science (AAAS). 1993. *Benchmarks for science literacy: A project 2061 report*. New York: Oxford University Press.

Dewey, J. 1913. *Interest and effort in education*. Boston: Houghton Mifflin.

Driver, R., E. Guesne, and A. Tiberghien. 2002. Children's ideas and the learning of science. In *Children's ideas in science*, eds. D. Rosalind, E. Guesne, and A. Tiberghien, 1–9. Philadelphia, PA: Open University Press.

Fennema, E., and P. L. Peterson. 1987. Effective teaching for girls and boys. In *Talks to teachers*, eds. D. Berliner and B. Rosenshine, 111–125. New York: Random House.

Gomez-Zwiep, S. 2008. Elementary teachers' understanding of students' science misconceptions: Implications for practice and teacher education. *Journal of Science Teacher Education* 19 (5): 437–454.

Kuhn, D., and E. Phelps. 1982. The development of problem-solving strategies. In *Advances in child development and behavior* 17, ed. H. Reese, 1–44. New York: Academic.

Kulkarni, S. 2009. Cowbirds in Love: 46. Death ray. *http://cowbirdsinlove.com/46*

Lunetta, V.N., A. Hofstein, and M.P. Clough. 2007. Learning and teaching in the school science laboratory: An analysis of research, theory, and practice. In *Handbook of research on science education*, eds. S.K. Abell and N.G. Lederman, 1st ed. London: Taylor & Francis.

Marek, E. A., S. J. Maier, and F. McCann. 2008. Assessing understanding of the learning cycle: The ULC. *Journal of Science Teacher Education* 19 (4): 375–389.

McComas, W. 2005. Laboratory instruction in the service of science teaching and learning. *The Science Teacher* 72 (7): 24–29.

National Research Council (NRC). 1996. *National science education standards*. Washington, DC: National Academy Press.

Nickerson, R. S. 1998. Confirmation bias: A ubiquitous phenomenon in many guises. *Review of General Psychology* 2 (2): 175–220.

Schauble, L. 1990. Belief revision in children: The role of prior knowledge and strategies for generating evidence. *Journal of Experimental Child Psychology* 49: 31–57.

Schauble, L., R. Glaser, K. Raghavan, and M. Reiner. 1991. Causal models and experimentation strategies in scientific reasoning. *Journal of the Learning Sciences* 1 (2): 201–238.

Schauble, L., L.E. Klopfer, and K. Raghavan. 1991. Students' transition from an engineering model to a science model of experimentation. *Journal of Research in Science Teaching* 28 (9): 859–882.

Tschirgi, J. E. 1980. Sensible reasoning: A hypothesis about hypotheses. *Child Development* 51 (1): 1–10.

CHAPTER 3

Using a Cycle to Find Solutions

Students Solve Local Issues With a Four-Step Cycle

By Glenn Fay Jr. //

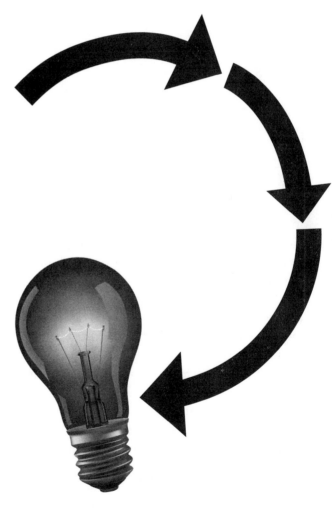

The National Research Council has suggested that science classes need to become more active and authentic (NRC 1996). To help with this effort, Thayer Model Engineering was created several years ago as an inquiry science course at Champlain Valley Union High School in Hinesburg, Vermont. The course is modeled after an introductory engineering course at Dartmouth College where the Thayer Model was developed, honoring one of the founders, Sylvanus Thayer (Frye 1996).

In this Champlain Valley Union High School course, students apply the Thayer Model—a four-step problem-solving cycle—to identify and solve problems using science, math, and technology. Students define a problem, describe specifications, determine a solution, and redefine the problem, which begins a new cycle. Through iteration, each cycle becomes narrower in focus and more clearly defines the best final solution to the original problem. Once students become familiar with the model, they create and enter in a competition a new technology designed with the intent to improve the world.

Thayer Model

In the course, students apply the Thayer Model to solve complex human problems. The model asks students to use their creativity to identify authentic needs and problems in their community and

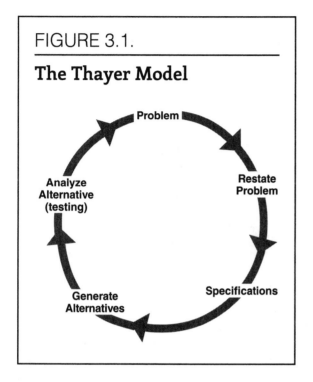

FIGURE 3.1.

The Thayer Model

Problem

Restate Problem

Specifications

Generate Alternatives

Analyze Alternative (testing)

in the world that can be served by applying science, math, and technology. These problems may be local issues, for example, related to pollution, transportation, security, safety, or health concerns.

The model uses four main steps in a cycle (Figure 3.1). First students *define a problem*. Once they have a problem, students describe *specifications* for the solution to the problem, which help define the parameters. With specifications in mind, students create *alternative solutions*, research and rate those different solutions, and choose a best solution. Students then *redefine the problem*, which begins a new four-step cycle. Fundamental to this model is iteration—each repetition of the cycle makes the individual steps more focused and brings students closer to the best, final solution. Depending on the product goal, students may work independently or in groups of two, three, or four. A case study applying the model is presented in this article.

A Case Study
Define a Problem
Students are presented with the following scenario: A cook finds evidence of mice in the school kitchen. First, students should identify the root problem and not just a side-effect that is symptomatic of a real issue. For example, if evidence such as mice droppings was the root problem, cleaning up the droppings would solve the problem. However, through discussion, students realize that droppings are just symptomatic of the actual issue—a mouse family, possibly with an expanding population. Once students have identified the root problem, the next step of the problem-solving cycle is to develop specifications, which can be used to define possible solutions.

Specifications
Specifications in the first iteration of the cycle should be general. The initial specifications, for example, might be that the solution must be safe, legal, ethical, effective, and economical. In the mouse example, this would mean that the specifications would not lead to solutions that use illegal approaches or could harm people, and that are affordable and actually solve the problem. General specifications allow students to begin to define what standards their solutions will need to meet. Specifications become more specific with each iteration of the problem-solving cycle.

Alternative Solutions
In the next step, students brainstorm and determine solutions that would meet the specifications. This step is where true creativity is needed. Because students are solving authentic problems with no certain viable solutions, sometimes the best resolutions are alternatives "outside the box." Alternative solutions that might meet the specifications of the mouse problem include using

poison, animal traps, high-frequency sound, or feline predators.

At this stage, students assign research roles to one another within each group to examine and learn more about the various alternative solutions. The roles ensure that the workload is distributed evenly and allows each student to become an expert in at least one area. This could include calling the National Rodent Patrol or conducting patent searches, for example. The different solutions are placed in a matrix to be cross-referenced with the specifications (Figure 3.2). Each solution is rated plus (+), minus (–), or neutral (0), based on the ability to meet each specification. The information gathered through research provides students with rationale to rate the potential effectiveness of each solution in the matrix.

This stage allows students to make connections as they apply what they learn. The information students gather is directly applicable to solving a real issue. This is a shift for many students and it results in heightened interest in learning. In the mice example—based on what students learned through their research and the results of the matrix rating—feline predators received the highest rating of the solution alternatives. Therefore, in this case felines provide the best solution to the mouse problem. However, the cycle does not stop here.

Redefining the Problem

With a best solution found for the problem, students in each group complete the first cycle by asking, "Is our problem solved? Or, do we have a new problem?" For example, feline predators were the solution decided on to correct the mouse problem. This decision creates new challenges and problems: What age and type of cat would be best for catching the mice? How many cats are needed? Students can pick one of the new problems to address, but more likely they will need to address all of the problems in order for the solution to be adequate. To address all problems, the overarching problem could be: Cats create care and maintenance problems.

The second iteration of the cycle might be based on the new problem: How do we maintain the cat population to solve our mouse problem? New specifications for the cat solution might

---

FIGURE 3.2.

Mice problem, first iteration: Solution–specification matrix

Solution	Safe	Legal	Ethical	Effective
poison	–	0	+	0
traps	+	+	+	0
sonic	+	+	+	0
felines	+	+	+	+

---

include physical health, efficient feeding and hygiene, mouse disposal, and cat storage. As the matrix in Figure 3.3 shows, contracting with a veterinarian might produce the best solution for solving the second problem of using cats to remove the rodents.

The second iteration gives rise to another, third problem—which veterinarian should they contract with in order to fulfill their best solution—which will lead to a new set of specifications and alternative solutions. After further research, and cross-referencing and rating the various veterinarians (alternative solutions), the best veterinarian (solution) is finally selected to work with. After multiple iterations, student engineers are able to develop ingenious solutions and the cycle becomes a familiar, almost intuitive process.

During the course, students have used the model to develop several prototypes—final solutions—including a new heating system for the school, a diesel exhaust scrubber to inexpensively remove carcinogenic soot from school bus exhaust, a robotic school hallway floor washer, a programmable medication dispenser, and a snowboard for paraplegic skiers. Each year, students generate a new crop of creative solutions

to complex human problems. Plenty of online resources exist for students to use during their projects, such as All Science Fair Projects at *www. all-science-fair-projects.com/resources.php.*

The Final Project

During the first quarter of the course, students are introduced to the Thayer Model with several smaller projects, such as the mouse problem, and they define technology and present their definitions to the class. Students discuss the range of applications of current technology. Next, the class studies the current state of emerging technologies, performs patent searches, and finally researches potential high school competitions. This brings students up-to-date with technology that is moving from conceptual stages into design and fabrication. Student awareness is expanded and they realize technology is farther along than they originally thought. Patent searches allow students to see if their own bright ideas have been patented or not. Each of these short units is scaffolding for the final project, which lasts all of second quarter. The course assignment for the second quarter is to create a new technology designed to improve the world. Students are then

FIGURE 3.3.

Mice problem, second iteration: Solution–specification matrix

Solution	Healthy	Feeding, hygiene	Mouse disposal	Cat storage
Veterinary contract	+	+	0	+
Custodial/maintenance	0	+	0	+
Student volunteers	0	+	0	0

required to enter their technology in a science competition.

Several possible high school science competitions exist, including NSTA/Toshiba ExploraVision, Thinkquest, and Siemens-Westinghouse. Most students choose the Exploravision competition because it demands that they develop a conceptual technology that is 20 years

Dartmouth/Thayer School Project

The Dartmouth Project for Teaching Engineering Problem Solving (online at *http://engineering. dartmouth.edu/teps*) provides teachers with a framework for bringing engineering problem solving into science, mathematics, and technology classrooms. Teachers use the Dartmouth/ Thayer School framework to guide their students through the problem-solving cycle and help them develop skills in communication, teamwork, and critical thinking. Students maintain high standards for scientific inquiry while they learn how to solve the less structured problems they will encounter in their future lives in classrooms and in employment.

The Problem-Solving Cycle

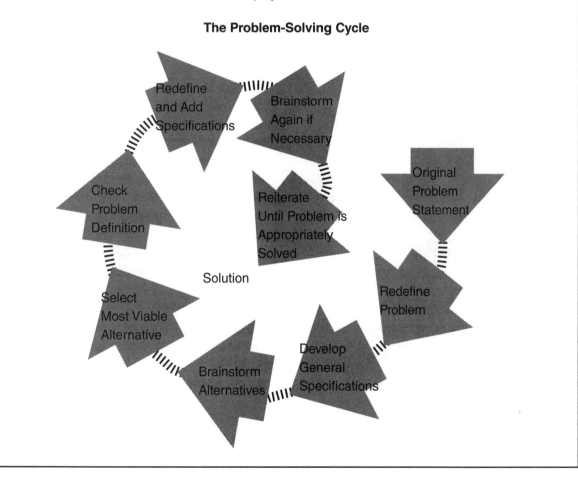

into the future—an ideal challenge with the Thayer Model—and the competition's timeline conveniently falls at the end of the second quarter.

ExploraVision requires students to describe the future technology, the history of the technology, breakthroughs that will need to occur for the technology to come to fruition, and consequences that may arise. The Thayer Model is the perfect process for identifying a human problem, finding a plausible future technology, and developing the best solution. Students work in groups of two to four to thoroughly research and develop a technology.

Another attractive aspect of the Toshiba competition is that it encourages students to work with an expert community advisor. In a world where scientific knowledge increases rapidly, even the best science teachers can become outmoded if they are the sole human resource for students. My students have worked with experts ranging from local cardiologists to neurologists, laser optics engineers, nanobot technologists, and computer programmers, and have developed e-mail relationships with researchers and specialists at universities around the world. These relationships have elevated the quality of thinking and student work.

Measuring Success

The Thayer course has been popular for several reasons. It teaches students an authentic problem-solving model that works with real problems. The model is versatile and students can see that it applies to personal as well as professional problems. [**Editor's note:** The Thayer website (see "Dartmouth/Thayer Project" sidebar, p. 17) includes an extensive archive of materials from teachers who have applied the Thayer model to different disciplines.] By the time students take this course, they often have acquired three or four years of traditional science content knowledge. The course allows students to apply that knowledge to authentic life situations. The pedagogy meets multiple learning styles. Many students who are not traditional learners, but nonetheless are hardworking and bright—possible future doctors, engineers, or science majors—can really do high-quality inquiry science using the background they have acquired in other classes. This is because students are put in the position of actively learning and applying their knowledge to creating new solutions to real problems. Students' inter- and intrapersonal, kinesthetic, spatial, and linguistic intelligences are valued and put to good use with project-based learning.

How do I know if the Thayer Model is worthwhile for students to learn and use? Aside from the anecdotal evidence and the glow of students excited about and deeply engaged in learning, several indicators show we are doing something right in this course. For one, the size of the class is growing. Word of mouth promotes this active learning class and attracts students who want to try new ideas. Over the past five years, we have enjoyed a national second-place Toshiba team, and three honorable-mention teams, all from a small rural school in Vermont. The bottom line is—students can use a simple but powerful problem-solving model to find solutions to complex professional and personal problems.

References

Frye, E. 1996. Engineering problem solving for mathematics, science, and technology education. *http://engineering.dartmouth.edu/teps/default_materials.html.*

National Research Council (NRC). 1996. *National science education standards.* Washington, DC: National Academies Press.

CHAPTER 4

Building Models to Better Understand the Importance of Cost Versus Safety in Engineering

By William Sumrall and Michael Mott //

Constructed in 1961, the I-35W bridge in Minneapolis, Minnesota, is the fifth-busiest bridge in the state. On August 1, 2007, it collapsed during rush-hour traffic, killing 13 people. According to authorities, its collapse was due to a design flaw involving a chemical reaction among salt, pavement, and steel, and unmet maintenance requirements.

On June 29, 2003, in Chicago, Illinois, 13 people died as the result of a balcony collapse caused by questionable material use and infrequency of inspections. Over seven years later, the cause of the collapse is still under litigation. Cited for poor construction, the owner is arguing that too much weight was put on the balcony.

While some disasters involving engineered structures are due to events in nature (e.g., tornadoes, hurricanes, earthquakes), others may be caused by inadequate materials, design flaws, and poor maintenance. These catastrophes result in the loss of human lives and cost billions of dollars. It is the responsibility of both public and private design and construction firms to ensure citizens are kept reasonably safe (under normal conditions) from physical harm when using roads, bridges, and buildings. Engineers and architects are given the primary responsibility of protecting the populace when construction is involved. Both governmental and private safety inspectors who usually have

backgrounds in engineering fields provide additional verification of a structure's safety. Engineers, architects, and builders have many factors to consider when planning for construction, including the moral and ethical obligation to protect fellow human beings, cost, and how long the structure will last and what kind of stresses it can undergo.

In the set of lessons described here, students design a tower, select building materials, determine building costs, and identify safety concerns. Working in groups with defined roles, students are actively engaged in project development, implementation, and evaluation. Modeling the work of engineers and architects provides relevance to the learning of science in the areas of physical mechanics and technological design and addresses National Science Education Standard F: Risks and Benefits for grades 5–8 (NRC 1996).

Build It Inexpensively, But Safely
Day 1
Tell students a guest speaker will be visiting the next day to discuss building structures such as bridges and towers. Have students brainstorm questions they have regarding safety and cost issues related to building of such structures. Record the questions on the board for discussion and editing. Questions students have come up with in the past include the following:

- How often are structures checked for safety?

- Who does the checking?

- Do private structures (e.g., balconies) ever get checked?

- Are there currently any structures in danger of collapse in the community?

- How do you guarantee a structure is sound and secure?

- How does the cost of materials factor into maintaining safety?

- What kind of stresses can various structures in the community hold?

- How long will various structures in the community last without maintenance?

After all the questions have been developed, ask students, working in groups of two or three, to investigate human-caused structural disasters (involving bridges, buildings, balconies, towers, stages, and so on) using the internet and local resources. The disasters students investigate should be due to human error or neglect. To investigate local disasters, you may need to give the groups an extra day or two to gather information. Figure 4.1 provides detailed directions for the activity.

Upon completion of the activity, have students share their research results. If a local disaster has been identified, you may consider a field trip to the site with appropriate supervision (e.g., fire marshal, assigned engineer). After they've completed their research, ask students if they have any more questions for the next day's class speaker.

Day 2
Bring in a local civil engineer, architect, or safety expert (e.g., from OSHA) to provide students with background information on the importance of safety and cost in the construction of bridges, buildings, and other structures.

Day 3

Begin the lesson by requiring students to do an individual search using the library or the internet to investigate historical structures. Ask students to identify and bring to class photographs or pictures of famous structures. Examples include the Eiffel Tower, Golden Gate Bridge, Arc de Triomphe, and Gateway Arch in St. Louis. Ask students to record observations about their structures. Some students will observe two-dimensional, geometric shapes within the structures. Some will readily identify triangles, squares, circles, and rectangles. As they continue to observe the images, point out students who have identified three-dimensional shapes (e.g., cylinders, pyramids, boxes, and so on).

After observing the images brought in by individual students, engage students in an activity to determine if these shapes are common to many structures and perhaps have a purpose. Ask students if they think the shapes have a purpose or if they are there solely for aesthetic reasons. Have students use plastic straws to construct geometric shapes for testing. Require students to use the same amount (e.g., one straw) of materials when comparing the multiple two-dimensional shapes they have created. Hence, each student can create one shape using one straw and create another shape using another straw. Ask students to develop a test to determine if shapes have different stabilities and strengths. One test students have used in the past involves setting their shapes upright and pressing downward with their index finger on the midpoint at the top of each figure. Through this activity, most students will determine that the triangle is a particularly sturdy structure. At this point, have students compare the strength of the triangle to the other figures that have been created. Ask students how downward forces are distributed (some should notice that the force is spread in two places at angles with the straw). Similarly, have students press down on the upright rectangle-shaped straw on its upper-length segment's midpoint. Ask students how the forces are distributed and which shape seems stronger and would be able to support the most weight. Additionally, have students draw out how they think forces are dis-

---

FIGURE 4.1.

Day 1 assignment

Using the internet, search for structural disasters caused by human error or neglect. Once you have located a disaster, let the teacher know so the rest of the class can be notified and there will not be any repeats. After the teacher gives you the OK to proceed, answer the following set of questions about the disaster your group located online.

- What was the nature of the disaster?
- What was the name of the structure?
- What human error caused the disaster to occur?
- What was the date of the disaster's occurrence?
- Were there any fatalities or injuries as a result of the disaster?
- How much cost was involved in repairing the structure, cleaning up, lawsuits, demolition, etc.?

Upon completion of this assignment, discuss with each other and parents the possibility of structural disasters (e.g., bridge collapses) in your area.

FIGURE 4.2.

How force is distributed with triangle- and rectangle-shaped straws

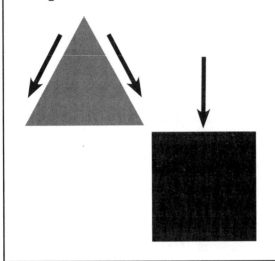

FIGURE 4.3.

Paper structures that hold significant mass

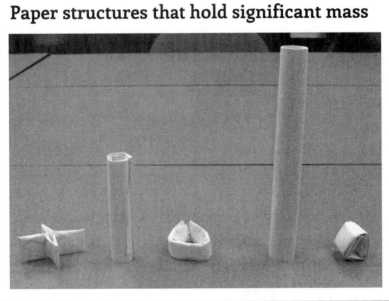

tributed using arrows on paper. Figure 4.2 shows how the force is distributed with a triangle- and a rectangle-shaped straw.

For homework, ask students to identify three two-dimensional and three three-dimensional shapes of local structures in the community. Dependent on school budget, you can take a field trip and have students take digital photographs of the structures.

Day 4

Begin class with a problem-solving activity involving the strength and stability of a piece of paper. Give students a sheet of paper (one per student) and ask them to shape it, without cutting or adding anything, so that it can hold a significant amount of weight. A large cup of pennies can be used for testing, though anything having a mass of 1 to 10 kg should be sufficient. It may take a while, but eventually some students will realize that rolling the paper into a cylindrical shape provides enough strength to support significant weight. Furthermore, some students will bend their cylinders to form a smaller but sturdier support if given enough time. There are several other possible shapes and folds that create significantly sturdy structures. Display the ones that are able to hold the cup of pennies and compare the structures for similarities and differences. Figure 4.3 shows some of the paper creations used to hold a large cup of pennies.

For homework, ask students to bring in for Day 5

"nature-made" structures that imitate these shapes (eggs, acorns, and empty bird nests are some of the items students have brought to class) or photos of harder-to-find items (e.g., spider webs).

Days 5–7

At the beginning of class, have students view the straw shapes, the paper shapes, and the natural structures. Ask them what they have learned in regard to strength of construction. Students should identify that triangles, arches, and cylinder shapes in particular provide sturdiness in the construction of structures.

In the next activity, students working in teams will build a "tower" using simple, inexpensive materials. Begin the lesson by creating teams of three or four students. If possible, arrange teams such that each group includes a student who is dexterous, one who is good at math, and one who has shown creativity in past problem-based assignments. Within the team, have group members choose their roles—engineer, construction company CEO, accountant, and architect. Hand out the activity worksheet and explain to students that their team is a construction company being asked to design and build a tower for a football field according to certain parameters, considering both cost and safety in its construction.

Explain that large projects are typically undertaken based on what is called a *sealed bid*. In the case of this activity, each company will be awarded the project for completion, since multiple towers will be needed for multiple football fields in a school district. However, students will still need to submit a sealed bid for review. To make things interesting, you (the teacher) will be judging project success based on two criteria: project cost and project safety. Furthermore, the company that builds the safest, least expensive tower will receive strong consideration when

---

FIGURE 4.4.

Materials and cost list

- Cable (thin wire): $20/cm
- Thin beam (wooden coffee stirrer): $100
- Wide beam (tongue depressor): $200
- Rod (plastic coffee stirrer): $50
- Welding (1 bottle of glue): $50
- Cement (4 cm modeling clay): $250
- Clamp (rubber band): $25
- Rod (wooden dowel): $25
- Planning: $50/minute
- Labor: $25/minute
- Maintenance/safety checks: $30/minute × frequency

---

bids go out to build a new, multimillion-dollar stadium. Point out that good planning in the long run may save teams money, and they will not be allowed any cost overruns without penalty.

Start the planning time clock for each company as soon as you have reviewed the activity worksheet and materials and cost list (Figure 4.4) with students and there are no questions as to what they are supposed to do. Emphasize that students cannot start building until their company has submitted a bid to the school district (i.e., you, the teacher). Thus, bids will be based on time to plan, time to build, and cost of materials. At a minimum, I allow two class periods for the construction of towers. While students are working, your primary job is to keep up with the planning, materials, and labor costs of each company, and to remind students of the real-world applications of the activity.

Building supplies (see Figure 4.4) can be purchased at most discount stores or in many cases rummaged out of supply closets. Plan to have 10 of each structural element (tongue depressors, stirrers, dowels) as well as a bottle of glue and one stick of modeling clay per group. Figure 4.5 is a photograph of inexpensive materials used by students in the development of their towers. Students should wear safety goggles during construction.

Day 8

On the last day of the project, students will test and measure their towers. Before testing begins, have each group show its tower to the class and ask them which elements of the tower make it safe, what makes it sturdy enough to handle high winds and rain, and how long they would

guarantee the tower's safety. Additionally, to continually focus on the importance of public safety, check to see if students considered a hand railing as well as some form of platform access (i.e., ladder) in their construction.

Quantifying the stability and strength of a tower is essential. Tests I have conducted in the past include the following:

- using a hair dryer at differing angles to determine tower stability;

- using a water spray bottle to simulate rain;

- doing strength tests over several weeks to determine durability (i.e., maintenance/safety check); and

- using incremental additions of weights to determine overall tower strength.

FIGURE 4.5.

Inexpensive tower materials

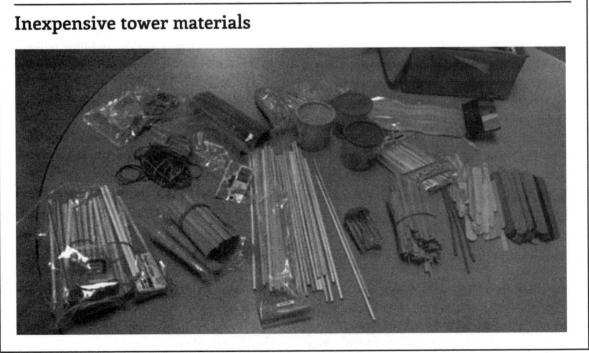

FIGURE 4.6.

Design, construction, and fiscal management rubric

Collaboration (step 1)	Tower qualities (step 2)	Fiscal considerations (step 3)
Superior Collaboration Team communicates effectively, solves problems, discusses options and possibilities, addresses different opinions articulated by peers, and resolves differences through discussion.	Tower strength is measured using incremental weights, wind, and rain testing. Through the teacher's testing in front of the classroom, tower strength is determined. A tower that holds the maximum weight dry, does not move when blown with a hair dryer at various angles, and holds the maximum weight wet is rated an A with regard to tower quality. An A+ rating is achieved if a company includes a railing and a ladder with its tower. If there were no collaboration problems (i.e., superior rating), then there are no deductions from the A or A+ tower quality rating. (Acceptable rating on collaboration drops tower quality rating to an A or B+, and an unacceptable rating on collaboration drops tower quality to B+ or B). Towers that receive A or A+ ratings move on to Step 3 and are assessed based on the fiscal considerations rubric section. (Towers rated lower than an A may be included for the fiscal considerations rubric section if only one company in the classroom receives the A or A+ rating. Usually, however, multiple groups (companies) receive an A or better rating.	Least expensive and safest (based on Step 2) tower is rated superior overall and the company is strongly encouraged to submit a bid for the new football stadium being planned. The A or A+ rating is maintained.

FIGURE 4.6 (continued).

Design, construction, and fiscal management rubric

Collaboration (step 1)	Tower qualities (step 2)	Fiscal considerations (step 3)
Acceptable collaboration Team communicates and solves some problems, discusses options and possibilities, and addresses different opinions articulated by peers, but does not fully resolve differences through discussion.	Tower quality testing fails one test. For example, it does not hold maximum weight dry (needs to hold at least half of incremental weights). Grade drops to a B+ in tower quality if collaboration is superior, drops to a B rating if collaboration is acceptable, and drops to a B- rating if collaboration is unacceptable.	Towers that do not go over bid budget maintain current A and A+ grade (usually only A and A+ towers are considered in the fiscal considerations rubric) and are considered acceptable towers, but possibly too expensive for future bid considerations.
Unacceptable collaboration Team communicates but does not solve problems to accomplish the goals of the project—one or two students' opinions are included and others' opinions and ideas are excluded.	Tower quality testing that fails two tests automatically drops to a B rating if collaboration is superior, drops to a B- rating if collaboration is acceptable, and drops to a C+ rating if collaboration is unacceptable.	Companies that go over on bids drop in rating by a full letter (A or A+ towers are dropped to B or B+).

To facilitate the testing, I suggest doing each test on all the towers before moving onto another test. For example, before moving on to another test, use the hair dryer at different angles on each tower and measure how far the tower moves or if it topples over. It is important to do the weight test in increments to differentiate which towers are the sturdiest. Students should create a table before beginning the testing for comparison of construction company data. Testing can be completed in less than 50 minutes.

Figure 4.6, p. 25, is a rubric I have used to grade projects. It is a three-step grading system where collaboration, tower qualities, and fiscal considerations are used to assign letter grades. Usually, more than one group achieves a superior on both collaboration and tower quality ratings of A+ or A, and fiscal considerations end up determining the winning tower project.

Conclusion

Building models while considering cost and safety together is a real-world science and engineering task that requires critical thinking and mirrors the professional approach of engineering and design teams. These activities involve students in recognizing the importance of what engineers, architects, construction workers, and accountants do, and show how much we depend on these professionals to make our world a secure place for travel and living. Students' initial investigations into disasters that occurred due to human error should be emphasized throughout the eight days of lessons.

An ethical dilemma is raised when students must weigh decisions relating to cost versus safety as they navigate the project, and the concepts of *safety* and *design* take on multilayered meanings.

A motivational and empowering conclusion for this project might involve student-led discussion and debate centered on these very ethical dilemmas resulting from the "cost versus safety" decisions: (1) What should be done when weighing the importance of cost to safety? (2) Who is responsible to ensure the structure is safe, who is responsible to ensure the structure is safe over time, who pays for it, and what is the source/rationale for their culpability? (3) What can we do to avoid structural disasters in the future? and (4) In the cost-versus-safety process, what tends to win out and why?

Reference

National Research Council (NRC). 1996. *National science education standards*. Washington, DC: National Academies Press.

Activity Worksheet: Build a Tower

Overview

Your team is a construction company being asked to design and build a tower for a football field according to certain parameters. You have two days to plan and construct your tower. Your company will need to submit a sealed bid before beginning construction of the tower. Because multiple towers will be needed for multiple football fields in the school district, each company will build a tower, and the cost and safety of your company's tower will be compared to other companies in the area. If the committee (your teacher) determines that your tower is the safest and most cheaply built, the committee will strongly encourage your company to submit a sealed bid for constructing a new football stadium. Quality work that is both safe and inexpensive can lead to greater rewards (i.e., additional contracts due to quality work) for a business. Similarly, work that is both unsafe and expensive lessens the chances of a business being profitable (i.e., no additional contracts due to shoddy work). Hence, the company that builds the safest, least expensive tower in this activity will receive strong consideration when bids go out to build a new, multimillion-dollar stadium.

Parameters

Towers must be a minimum of 15 cm in height and have a square base that is a minimum of 5 cm, and each tower should have a platform to support a football coaching staff. A ladder and handrails for safety will be judged favorably.

Materials you can use and the prices for materials and labor are listed below (see Fig-

ure 4, p. 23). Keep in mind that good planning at the outset may save your company money in the long run. Your company will be penalized for cost overruns—additional labor or materials costs beyond the budgeted amount could result in the company's bankruptcy and possible subsequent lawsuit for your default on services. If a project does overrun in cost, this will be factored into the assessment and determination as to which company will be encouraged to submit a bid for the new stadium.

Planning time cost is part of the budget. Thus, the clock will start as soon as you have reviewed this sheet and there are no questions. Remember, you cannot start building until your company has submitted a bid to the school district (i.e., the teacher).

Team jobs

Architect: You are the leader in the planning stage and will determine how the tower looks. In collaboration with the engineer, you will need to be thinking about safety at all times. In collaboration with the accountant, you will need to consider cost. In collaboration with the construction company CEO, you will need to consider both cost and safety.

Construction company CEO: Your primary job is to make sure the tower is built. In collaboration with the engineer, you will need to be thinking about safety at all times. In collaboration with the accountant, you will need to consider cost. In collaboration with the architect, you will need to consider both cost and safety.

Activity Worksheet: Build a Tower (continued).

Accountant: Your primary job is to make sure construction costs are no more than the sealed bid. In collaboration with the engineer, you will need to be thinking about safety and cost. In collaboration with the construction company CEO, you will need to think about cost. In collaboration with the architect, you will need to consider both cost and safety.

Engineer: Your primary job is to make sure the tower is safe. In collaboration with the architect, you will need to be thinking about safety at all times. In collaboration with the accountant, you will need to think about safety and cost. In collaboration with the construction company CEO, you will need to consider both cost and safety.

Sealed bid
(to be submitted to the teacher)
Students:
Closed bid amount:
Labor cost breakdown:
Materials cost breakdown:

CHAPTER 5
Miniature Sleds, Go, Go, Go!
Engineering Project Teaches Kindergartners Design Technology

By Gina A. Sarow ///

Imagine kindergartners eagerly creating blueprints for their own inventions. Imagine them actually building them with tools. Now imagine them testing out their design and explaining the science behind it to their peers! Not possible, you say? It is with *Design Technology: Children's Engineering* (Dunn and Larson 1990), a learning model I implemented in my classroom two years ago. This model allows children to experience developmentally appropriate, hands-on instructional methods that enable them to draw, plan, design, build, test, and improve their solutions by applying their knowledge to new and different situations.

Engineering and design technology is not a separate subject but rather a supplement to classroom lessons. In my classroom, I began incorporating one design technology lesson each month to coincide with such themes as plants, bugs, and the weather, and it worked beautifully. We always began with a related children's literature story—thanks to our wonderful librarians—and branched out to other areas of curriculum, such as science, mathematics, writing, and art.

Examples of engineering projects I've done with students include building a sailboat that sails in a wind channel, building a bug catcher to take home and catch live bugs, and designing a snake that moves on both ends. One of the most successful engineering and design activities I've

conducted with students was a miniature sled-building activity that resulted from a brainstorming session among two teachers and myself.

A Sledding Party

The sledding project began after a huge snowfall (20–25 cm) one November. I took the children to the library and read aloud *The 14 Forest Mice and the Winter Sledding Day* by Kazuo Iwamura (1991), which relates how a father mouse kept his little mice busy during the long snow season by planning, drawing, and building sleds. In our Wisconsin community, the wind chill often makes it feel as though the temperature is below freezing, so children usually have to stay indoors. I challenged the children to build a miniature sled that would go down a hill carrying weight—this was our way of keeping busy like the mice.

Soon after reading the story, I sent home a newsletter about the sled activity and asked parents to help their children bring their sleds to school for a sledding party the next day. (Sleds are not allowed on school buses for safety reasons.) Parents brought in an assortment of sleds, including cardboard sleds, wooden toboggans, and plastic sleds and saucers. One child's mother even brought in a catalog flier showing different types of sleds on sale at a local store, and another parent contributed an article from the newspaper about what makes a good sled.

It was a beautiful day for sledding: The sun was out, the air was warm, and the snow was tightly packed from snowmobiles traveling to school. After an afternoon of sledding, the children and I sipped hot chocolate with marshmallows, which provided a great opportunity for the children to talk about what materials worked best for sledding and why. They discussed their sleds in groups of two so that they could bounce ideas off each other, an important concept when beginning open-ended lessons for kindergarten children.

The children reached a consensus that a "good sled" was one that moves fast. Some children thought metal would make the sleds go the fastest because snowmobiles have metal skis (many students have older siblings who drive snowmobiles to school). One child said the speed could come from the same power—weight and gravity—we used in building fire trucks (a previous engineering project). Others discussed using plastic for their sleds because many of them used plastic sleds and saucers at home.

Classroom Construction Site

Materials for the sleds and the space to build them came from our classroom construction site, where students regularly practiced with tools and made their own designs. At the beginning of the year, I sought help from parents and other volunteers to build a construction center in our classroom, supplied with recycled materials of every shape and kind for projects. I sent home an initial newsletter asking parents and relatives to save various boxes, paper towel tubes, plastic foam, thread spools, milk cartons, egg cartons, fabric, milk caps, plastic caps from peanut butter jars, buttons, wood scraps, duct tape, and aluminum pie tins. Parents were enthusiastic about the construction center and contributed a wide range of materials, including oatmeal boxes, wheels and gears from nearby factories, and colored wire from the local phone company. Parent volunteers organized these materials in plastic stocking shelves in the classroom.

Students used this construction center throughout the year for all of their engineering and design projects. I purchased tools that were just the right size for very young children using funds from the Herb Kohl Educational Foundation. (I was a Kohl Teacher Fellowship recipient.) Criteria for tool size was based on the children's sizes and their experience using such tools.

Tool Safety

Parents also volunteered to teach tool safety. During class visits, they showed the children various tools, taught the correct name for each tool, and demonstrated how each tool was used in an appropriate manner. During these sessions the parents practiced building with the children. While making a simple hat hanger, for example, parents showed the class how high to swing a hammer (never higher than your shoulder for control) and how to use the reamer (hand tool for enlarging 3/16" holes).

The construction center enabled students to develop and practice the skills required for construction. Many teachers, administrators, and parents may feel apprehensive at the thought of young children using such tools as small hammers, hole punchers, and vices. I feel, however, that if children are given the appropriate size tools for their small hands and shown how to use them correctly, they quickly develop the fine motor skills required to use them safely (Idle 1991).

The construction center also served as a great way to involve parents. Since we began engineering and design projects, parents who initially came in only for scheduled meetings were now visiting the classroom weekly to see their children's constructions. As their children led them to their creations, I realized that the students had told their parents a great deal about their work and this energy was spreading!

Let's Get to Work!

On the day after the sledding party, the children rummaged through the stocking shelves of the classroom construction center to gather materials to make miniature sleds. After examining the materials, they sat down with their partners and drew a blueprint of their design. (See Figure

5.1.) Students understood blueprints as "a plan to help make something." They had learned to work with blueprints earlier in the year using *Sammy's Science House* CD-ROM from Edmark (1996), in which users choose a blueprint and click on materials to place, and *Tigger's Contraptions* CD-ROM from Disney (1997), where users pick out images of recycled materials and assemble them with a mouse.

FIGURE 5.1.

Student blueprint examples

Our Sled Blueprint

Partners: Colton sise ily

Challenge: To build a sled that will slide down a hill of snow while carrying weight.

Partners: who and Jer py

Challenge: To build a sled that will slide down a hill of snow while carrying weight.

The children began to build their sleds on the third day in our classroom's construction center. They worked on their sleds every day during free time for two weeks. With adult supervision, some groups used a junior hacksaw to cut some pieces of wood, and one group used a junior hand drill (all of these tools were manual tools, not power) to make holes to insert miniature flag poles. Most groups used duct tape and masking tape for the sleds' final touches.

While I walked around the room to listen to their discussions, I noticed one group wanted to build their sled in the form of a triangle because both students liked the shape. They did not yet realize that the shape itself is a hindrance for sliding. I asked the group and surrounding students, "Do you think the shape of your sled will help it move smoothly through the snow?" and "Do you have an idea of how to build the skis so the shape does not matter?" Some children interjected that the points of the triangle would get caught unless they use some type of ski system. Other students

This team considers adding a long handle to their sled as they draw a blueprint.

suggested bending the cardboard to make the sled go down the hill easier, otherwise the points would get caught in the snow.

"Kid Watching"

I am always amazed at the unexpected ways in which children add to and improve their blueprints. During this stage of the creation process, I rarely interfere with the students' designs to avoid directing them to one "right" solution. Instead, I aim to develop the child's thinking, reasoning, and problem-solving skills, which has taught me to give up on the "quiet" classroom. This is a time for the children to talk, share, disagree, and create opportunities to learn. Be prepared for a high level of noise with this type of interaction.

I used my "kid watching" skills to assess learning. Kid watching occurs when I stand back in the classroom and observe the children from a distance. I watch and listen to how they interact with each other, how they share tools, use their manners, and how they accept challenges. I also ask searching questions. These types of questions usually develop when I have the patience and courage to stand back and give them the time necessary to accomplish their goals. Children enjoy having opportunities to figure things out on their own and with the help of peers. As a teacher, the two questions I ask most are "How did you get that?" and "Why did you do it that way?" For this particular activity, most of the children's responses to my questions came from their past experiences, such as what they had observed with sleds and snowmobiles.

Sharing Knowledge

On the fourth day, the children stood in front of their kindergarten classmates and explained how their sleds were designed and built and how they thought their sleds would work in the snow.

Some groups explained how they chose a design just by the "looks" and the color. Other students thought the whole concept out right down to the handle with which to carry the sled up the hill.

During the presentations, some groups' comments about their sleds included

- "We used a plastic-foam egg carton because it had 12 seats (holes), and our friends could sled together. The clothespins help keep the people in, and they act as a bumper in case we get close to the trees. Our skis are wood. We made a long handle so we could drag it up the hill for another ride"; and

- "We used cardboard because when Joe goes sledding he uses cardboard boxes and they go fast! We added a steering wheel so we can steer away from other sleds and trees. Our sled is covered so we do not get wet when it is snowing hard—it hurts my face when it snows really hard."

In their presentations, the children used engineering terminology correctly. I taught them earlier in the year that using "engineering words" would help them understand one another's ideas more clearly. In the sled presentations, for example, instead of saying "our sled needs to be bigger," they used the words *taller*, *wider*, and *heavier*. Those words helped with planning and building the sleds.

The sled-building activity allowed the children to use all the information learned in previous design projects about levers, linkages, and pulleys. (In previous lessons, they used a pulley system to build a spider that could move up a web, a lever to create a turtle whose head moved in and out of its shell, and a gear system to make a simple clock.)

For example, one group—incorporating what they had learned about pulleys—attached a string

"Our skis are wood," explains one group as they present their sled to the class.

to their sled with a weight at the end. They tested it by placing the sled at the top of the classroom's indoor slide and letting the weight go. The sled worked on the smooth surface of the slide but did not function outdoors because of the snow. Another group knew from a previous lesson how to use a compass to draw a circle and then cut out a cone. They used this knowledge to make cones on their sleds to "guard your face from the snow."

Testing Celebration

The fifth day was a day of celebration! The children couldn't wait to "test" their designs out in the snow. Many of the children's parents came to school that day to see the sleds.

The students used weights ranging from 0.45 kg to 2.25 kg to put on the sleds. (We made the weights by filling plastic bags with small pebbles until each weight increment was reached.) Since the children's definition of a good sled was speed, all the sleds worked in some way. However, their idea about a good sled changed as they realized that just getting the sled down the hill presented a challenge. The sleds made of cardboard proved the fastest. I asked their creators, "Why is it so fast?" They figured it had to do with the curve of the cardboard. The sleds with skis made it halfway down the hill but then stopped when they reached the bottom. Two sleds did not work at all because the weights kept sliding off. The children soon realized that they forgot to secure the weights to the sleds (the plastic bag became very brittle and slippery outside).

The children's reactions to their sleds were mixed—some felt disappointed that their sleds did not make it off the top of the hill, some cheered as the sleds completed the run, and others wanted to go to a taller hill to see if height made a difference.

After testing the sleds for about 20 minutes, we went inside to discuss what worked well and what we could change. From their previous design activities, students had already realized that their creations don't always work the first time; sometimes we need to go back to "tinker" with a project to make it better.

Engineering Pluses

According to *Still More Activities That Teach* by Tom Jackson (2000), "for children entering the first grade in 1997, 50 percent of the jobs that they will have in their lifetime have not yet been invented." So how do you prepare children for those jobs? Jackson stresses the three Ts: technology, thinking skills, and team cooperation. I believe these life skills can be developed through engineering and design projects. The main goal in my projects is exposure—I want kindergarten students to begin exploring, problem solving, drawing blueprints, practicing tool safety, and working together with partners in cooperative learning groups.

Kindergarten children already believe that learning is fun, easy, and exciting. They also feel that they can do anything. These hands-on, action-oriented problem solvers are walking, talking engineers just ready to explore, build, and discover. Our job as teachers is to keep their playful attitude alive. Kindergarten rooms incorporate more time for play; as children get older, this time quickly vanishes. As teachers we strive to keep this enjoyment and excitement for learning alive; engineering and design technology activities can help you do this. It has worked in our room—give it a try!

References

Costa, A., and R. Liebmann. 1997. *Envisioning process as content: Toward a Renaissance curriculum.* Thousand Oaks, Ca.: Corwin.

Dunn, S., and R. Larson. 1990. *Design technology: Children's engineering.* Bristol, PA: The Flamer Press.

Idle, I. 1991. *Hands-on technology.* Cheltenham, England: Stanley Thornes.

Jackson, T. 2000. *Still more activities that teach.* Salt Lake City, UT: Red Rock.

Sigmon, J.F. 1997. Children's engineering: The use of design briefs. *The Technology Teacher* (56): 14–24.

Children's Literature

Iwamura, K. 1991. *The 14 forest mice and the winter sledding day.* Milwaukee, Wisc.: Gareth Stevens.

Software

Ready for MATH with *Pooh: Tigger's contraptions.* 1997. Burbank, CA: Disney Interactive.

Sammy's science house: The workshop. 1996. Redmond, WA: Edmark.

Internet Resources

Mrs. Henriksen's Poetry Pages. *members.home.net/henriksent*

CHAPTER 6
Science and Engineering

By Donna R. Sterling //

When teaching the scientific principles of flight, I have students create design prototypes, engineering projects that are initial attempts to solve a problem. They require students to apply their knowledge of science in order to construct a working model. In the process of designing and testing working models, students grapple with and develop an understanding of basic principles of science and engineering technology.

From a teacher's perspective, design prototypes are one-page handouts based on engineering protocols (see Design Prototype: Flight Distance, p. 42). These easy-to-use handouts organize instruction while engaging students in the design and construction of projects such as airplanes, cars, boats, buildings, filtration systems, and flashlights.

Engineering Protocol

The handout is broken into three parts. At the top is what's referred to as a "white paper," which outlines the context of the problem, need for a solution, and proposed engineering project to solve the problem. In the middle are the design specifications, which outline the specific construction and performance criteria that the project is to meet. At the bottom are the company (classroom) resources—a list of the human resources, equipment, materials, energy, and capital that are available. It also specifies the timeline to be followed and the reporting criteria. The design prototype format adapts systems engineering principles and closely parallels a process known in industry as IRD (pronounced eye-rad), which stands for Industry Research and Development. This real-world connection should be discussed to lend an air of authenticity to the projects.

Starting the Lesson

Start the lesson by having students identify what they know and want to know about flying. As a natural part of this discussion, students identify that they want to know how things fly. To further set the stage for introducing the design prototypes for flight distance, ask students to identify the first people to successfully fly. The students typically know about Orville and Wilbur Wright. Then inform them that they will be investigating flight just as Orville and Wilbur Wright did in the late 1800s and early 1900s and as aeronautical engineers do today.

Next, hand out the design prototypes on flight distance for students to read (p. 42). After five minutes, answer any questions that they have. If design prototypes are new to the students, explain the IRD process and what each part of the handout simulates. Then distribute the materials for constructing the planes. The room will soon become a hubbub of activity as the students construct, test, redesign, and retest their planes. Once they are engaged, it is time to introduce the principles of flight and have students apply them to their preliminary investigations.

The Principles of Flight

To focus students on the principles of flight, I interrupt them after about 20 minutes of design and testing and ask them "What are the forces acting on a plane?" To guide their thinking and discussion, I use a transparency with a picture of an airplane

on it and the question written out (see Figure 6.1). I find that students can usually identify the four forces (lift, weight, thrust, and drag), but have an incomplete understanding of the concepts (see Figure 6.2). For example, they may not understand the relationship between gravity and weight, so a guided discussion can clarify the relationship. After all four forces have been identified, I show a second transparency that labels the directions of the four forces (see Figure 6.3).

To deepen their understanding, the students complete a chart comparing how to change the four forces acting on airplanes (see Figure 6.4, p. 40). They first work individually and then in pairs to complete the chart. During the discussions, I circulate among the groups to listen to their explanations and help clarify their understanding.

Next they draw diagrams, first individually and then in pairs, to illustrate the principles

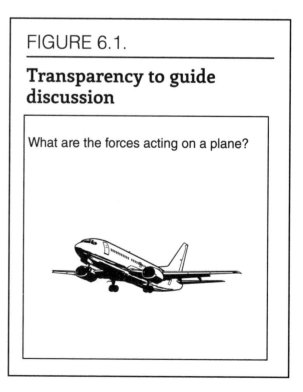

FIGURE 6.1.

Transparency to guide discussion

What are the forces acting on a plane?

of drag and lift. To explore their drawings and thinking further, I have students work in groups of four to discuss and refine their diagrams. I then ask the groups to take turns sketching their diagrams on the board and explaining how the prototype designs can be modified to increase or decrease the forces of drag and lift.

After the presentations, we discuss the diagrams as a class. Depending on the accuracy of the diagrams, the discussion can proceed in many directions. I channel the discussion to identify or create at least one accurate drawing showing what causes drag and lift. Because the concept of lift is more abstract, further reading, discussion, and experimentation may be required before an accurate diagram can be created. My goal is to have students understand that the difference in pressure below and above the wing creates lift. It is up to students at this point to decide how best to generate this difference in their design prototypes.

At the conclusion of these discussions, the students return to designing and testing their airplane prototypes. If the discussions have been successful, the students' chatter is now littered with their new scientific vocabulary as they continue designing and testing their airplanes. Students try to reduce drag by bringing the front of their planes to a point or they may try to increase the lift generated by the wings by changing the thickness and curvature.

Performance Tests

I always hold off on flight performance tests until the second day, because many of students continue the design and testing process at home. Some even decide to hold a competition at home to test the designs of other family members.

Finding an appropriate location to conduct the flight tests is critical to ensure the accuracy of the science. I have successfully held the tests in

my classroom, the hall, and the cafeteria. Testing planes outdoors is not a good idea because they are susceptible to even gentle breezes. A large area clear of obstacles where many students can test their plane simultaneously is ideal.

FIGURE 6.2.

Forces acting on the plane

Lift is the force that must overcome the weight of the aircraft.

Weight is the force of gravity on an object.

Thrust is a force created to push an object forward.

Drag is a force that resists the motion of an object due to friction and pressure differences.

FIGURE 6.3.

Direction of forces acting on a plane

FIGURE 6.4.

Guided student inquiry

How do you control the forces acting on an airplane?

Force	To fly, do you want to increase or decrease this force?	How can you accomplish this— increase or decrease?
Thrust		
Drag		
Weight		
Lift		

Draw a diagram explaining the scientific principles involved in changing drag.

Draw a diagram explaining the scientific principles involved in changing lift.

I mark off the floor in 1-meter increments from the starting line to a distance of 10 meters. Everyone who gets his or her plane to fly for 10 meters is a winner, which is most everyone. Of course, students will probably insist on conducting their own unofficial contest to see whose plane travels the furthest.

Before the flight tests, I remind students to write their name on their plane. Depending on the size of the space, I try to have six or seven students test their planes at the same time. They line up at the starting line and release their planes on the count of three. All students whose plane did not go 10 meters are given two additional chances to cover the required distance.

Assessment

I print the assessment rubric (Figure 6.5) on the back of the design prototype. This way, students know the performance criteria from the beginning. Adjustments are made to the rubric based on the special needs of students in the class. For example, the English section would allow for more errors for students who are still learning the language.

Science Standards

Both the National Science Education Standards and Benchmarks for Science Literacy call for a student-centered learning environment that actively engages students in asking questions and designing experiments to solve problems (NRC 1996; AAAS 1993).Furthermore, science and technology is one of the National Science Education Standards at all grade levels. According to the national standards, "The relationship between science and technology is so close that any presentation of science without developing an understanding of technology would portray an inaccurate picture of science."

Additionally, the science and technology standards offer students "experience in making models of useful things, and introduce them to laws of nature through their understanding of how technological objects and systems work." "In the middle-school years, students' work with scientific investigations can be complemented by activities that are meant to meet a human need, solve a human problem, or develop a product." The standards emphasize the students' abilities to design solutions to problems and the relationship between science and technology.

Design prototypes are an easy yet effective strategy to actively engage students in their own learning. They are hands-on, inquiry-based projects that focus students on designing their own solutions to problem situations similar to those encountered by scientists and engineers.

FIGURE 6.5.

Flight distance rubric

	0	2	4	6	8
PERFORMANCE					
Paper	nonstandard	–	–	–	1 piece of paper 216 × 279 mm
Launch	variation	–	–	–	one hand only
Landing	on nose and and upside down	–	on nose or upside down	–	right side up
Trials	>3	–	–	–	≤3
Distance	<7 m	7 m	8 m	9 m	≥10 m
Meets all critera	no	–	–	–	meets all above maxium criteria
REPORTING					
Model	missing	falling out of notebook	loosely attached in notebook	firmly attached and partially labeled in notebook	firmly attached and labeled in notebook
Drawings	missing	not clearly drawn; successful and unsuccessful models	not clearly-drawn; successful or unsuccessful models	clearly drawn; successful or unsuccessful models with explanations	clearly drawn; successful and unsuccessful models with explanations
Writing	missing	very difficult to follow	unclear and rambling	unclear or rambling	clear, concise, and accurate
Science	missing	≥2 science misconceptions	1 science misconception	minimal accurate science principles	lots of accurate science principles
English	needs major improvement	5–6 spelling and grammar mistakes	3–4 spelling and grammar mistakes	1–2 spelling and grammar mistakes	no spelling or grammar mistakes

References

National Research Council (NRC). 1996. *National science education standards*. Washington, DC: National Academies Press.

American Association for the Advancement of Science (AAAS). 1993. *Benchmarks for science literacy*. New York: Oxford University Press.

Resources

Collins, J.M. 1989. *The gliding flight: Twenty excellent fold and fly paper airplanes*. Berkeley, CA: Ten Speed Press.

Klutz Press. 1998. *The best paper airplanes you'll ever fly*. Palo Alto, CA: Klutz Press.

Design Prototype: Flight Distance

Context

Over the years, people such as Orville and Wilbur Wright have attempted to solve the mysteries of flight.

Need

As part of the design process people have built small-scale models that they have attempted to fly. They have experimented with the notion that a flying machine can be made by folding a piece of paper at various angles.

Design brief

Design and construct a flying vehicle that is capable of unassisted flight over a distance of 10 meters.

Specifications

1. The vehicle will only consist of one piece of 216 mm by 279 mm paper.
2. The flying vehicle must be capable of unassisted flight over a distance of 10 meters.
3. Thrust will be provided by one hand.
4. The vehicle may not land directly on its "nose" or upside down.
5. Designers will be given three attempts at the designated time to achieve the desired results.

Resources

People—You will work by yourself as the designer, fabricator, and evaluator.

Tools/Machines—There are no tools or machines available to work with.

Materials—One piece of 216 mm by 279 mm (8½ × 11 in.) paper.

Energy—You may use the energy generated by your prolific brain cells and the thrust provided by one hand.

Capital—You have absolutely no capital (money) to work with.

Time

You have the remainder of this science class and the first 15 minutes of the next science class to design, construct, evaluate, redesign, and evaluate your vehicle prior to the scheduled performance test.

Reporting

Put your best working model and drawings of successful and unsuccessful designs in your science notebook. Explain why the different models worked and didn't work.

CHAPTER 7

Elementary Design Challenges
Students Emulate NASA Engineers

By Jonathan W. Gerlach ///

How many of our students come to the classroom with little background knowledge about the world around them and how things work? If students' hands-on experiences only scratch the surface of true understanding, they'll be left with more questions than answers. To help students develop conceptual understanding and explore the design process, I brought the NASA "Engineering Design Challenges" program to my district, redeveloped for elementary students (see Internet Resources). Our program trains teachers how to use real-world scenarios to teach scientific concepts through engineering design challenges.

At "Educators at Space Academy," I trained and completed the challenge and process myself, and a lightbulb clicked. Students need more than one chance to be successful at a task. So many times they are left thinking, "next time I would have…." The design process allows students to have that next time.

I put this program into practice in some of our renaissance Title I schools in which students seem to lack the most background knowledge when it comes to science. In a fifth-grade classroom, I taught a lesson on forces and motion while discussing how more than one force affects an object at the same time. I set the stage by telling students that they were aerospace engineers working for NASA. It was then that students began to see how science is used outside the four walls of the classroom.

Engagement Activity

The class was engaged by a video compilation I put together of various types of planes and flying machines (see Internet Resources). (A simulation, PowerPoint presentation, or game could also be used.) This video gave them a background on what NASA has already developed, so they knew where we (NASA) had already been. I also showed them a flash animation from NASA all about X-planes through history (see Internet Resources). In general, the engagement activity should be based on a real-world scenario to build understanding of how science is used in the real world (e.g., plane design, levee design, lunar lander, or shipping container design).

Introduce the Design Process

The design process is a learning cycle that allows students to solve problems the way scientists and engineers do in the real world. Engineers use a continual process of design–build–test–record data–analyze data and redesign until they have

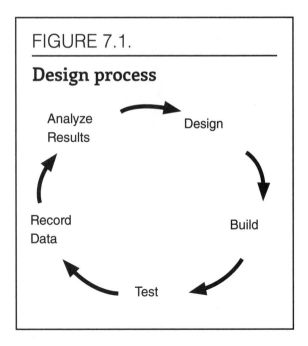

FIGURE 7.1.

Design process

Analyze Results

Design

Record Data

Build

Test

created a product that meets their specifications. Some engineers continue this cycle for years to solve a problem. To help students visualize the cyclic nature of the design process, I have provided a chart that you can use to lead class discussion and stimulate accountable talk (Figure 7.1). By having students communicate thoughts, ideas, and theories to each other and myself, students are learning the importance of communication to a scientist and they are held accountable for their own learning. Posting the design process in the classroom is a great reinforcement tool to remind students that the process is just as important as the result. The design process can also be used to teach the importance of models and how two- and three-dimensional models can be used to evaluate and discuss creations before the final outcome. During this activity, I asked students to think about how engineers might work on a product. We talked about drawing blueprints, conducting trials, and going back to the drawing board. I then showed the students the design process chart (Figure 7.1) and had them copy it into their notebooks.

Introduce the Design Challenge

Students should be first placed in groups of two. With this activity, too many students working on the same X-plane can make it hard to see who understands what. The teacher can set this design challenge as a contest in which students think they are competing against each other, but ultimately they are competing against themselves. Growth and seeing the learning process is what is being looked for as the ultimate goal. I challenged students to create a plane from Styrofoam and paper clips that would fly the farthest. They have limited materials, just like real scientists on a budget, and are only limited by guidelines set by their superiors at NASA.

In this challenge, the students will build an X-plane from provided materials that will glide through the air for the farthest distance in a straight line from the initial start point. Students first measure the distance they can get a paper airplane to fly. Then they design, build, test, and revise their own X-planes. Materials include the following: plastic foam food tray (approximately 28 cm × 23 cm; size 12, supermarkets usually pack meat on these), paper clips (two jumbo paper clips, two regular-size paper clips), sharp pencil, plastic knife, scissors, toothpicks (with rounded edge), and goggles (for eye protection). When students are working with sharp objects (e.g., scissors, plastic knives) precautions and laboratory safety rules should be reviewed and followed carefully.

This challenge is open-ended and inquiry-based. I introduced the challenge and let the students begin working. Through watching the students work, I was able to see what kind of background knowledge they have in this subject. For example, a lot of students immediately built planes with small wings and no tail section, like a paper airplane. When they tested the plane, I asked them to think about what a real airplane looks like and how their plane is different than a real plane. I was also able to assess and spot misconceptions and plan instruction accordingly. Formal instruction is provided after the students test their X-planes the first time, before I have them go

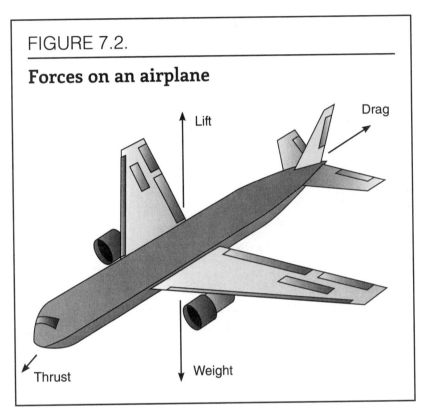

FIGURE 7.2.

Forces on an airplane

Lift

Drag

Thrust

Weight

back and redesign, starting the design process over. This way they are able to learn through experiences and the formal instruction clears up misconceptions and introduces vocabulary.

Many of the students in this class had never been on an airplane, had little background knowledge about planes, and had many misconceptions about how things fly. For example, many students thought planes had to go fast to fly, and they must look like jets.

Build and Test Design

Each group is given graphing paper and asked to draw a "blueprint" for its plane's design, based on how students want to build it. No real parameters are given to the students other than each group's design must be explained before building commences.

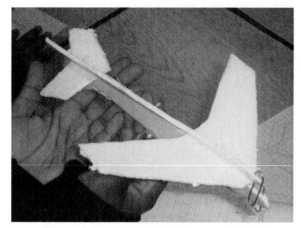

A student's plane is ready to fly.

Students used an accountable talk strategy I call P B & J (Pivot, Blab, and Jot) in which they talked to a neighbor about similarities and differences in the planes they've designed, and then wrote in their journals about their discussions. Students have had experience writing in journals prior to the design challenge.

Students were allowed to change their design as they tested their plane as long as they changed their blueprint as well. As students were building and testing their planes, they became lost and frustrated at times because of the lack of background knowledge they possessed. When asked questions about their planes, students explained that the designs were mostly based on cosmetic appearances (instead of an understanding of flight). They spent extra time adding racing stripes and color, instead of changing their plane design after test flights. When asked why their plane flew a certain way, they were unable to explain their thoughts. This was due to a lack of understanding of the concepts and a lack of command of terminology.

Students are often left with that feeling of wanting to try again, knowing they can do better. Before the students begin again, discuss as a class the different design successes and failures that occurred. Use scientific/engineering terminology with the students and discuss any misconceptions they have. Focus on the concept you are trying to teach. With the X-planes, we discussed balanced forces and unbalanced forces. I drew an example of a plane on the board and had students do the same in their journals. As a group we then discussed what forces might be acting on the plane and how we use the forces to make the plane do what we wanted (Figure 7.2, p. 45). We brainstormed ideas on how to increase thrust and lift. Students always wonder why their plane seems to only do flips. This is a time to create talk about how the mass of the plane must be level to fly in a straight line. One student said, "Oh, that's what the paper clips are for...." At this point, take all the materials away, and have every team start from square one.

Modify Product

This time, as students start from square one, they will have an actual experience to base their designs on. They will have knowledge of terminology and have a beginning understanding of the concepts that may lead to success. The students were unable to do this the first time, which led to frustration, because they had no experience to look back on. The students were taking stabs in the dark or trying to design something based off misconceptions.

Each group's design is tested and recorded by the entire class. Each group discusses their ideas and design concepts before its product is tested. When students went back to the drawing board, they worked with a renewed vigor; it took them half the time to design and build their planes. Because they now had an experience to base their design on, they knew what worked, they knew what didn't work, and they had a basic understanding of the physics of flight. They

46

started to use the terminology in their explanations and showed an understanding of balanced and unbalanced forces.

When students had to explain their final product to the class before the final test flight, there was a miraculous change in the conversation. Students were using physics to explain their designs instead of what "looked cool" and were informally assessed based on these descriptions. One group described their giant wings and weight distribution as "just like a Frisbee, the weight has to be equal everywhere...." Another group said, "...we used larger wings to allow more area for more lift...." Assessment for this lesson can be done through anecdotal notes, journal checks, and conferencing with students. The discussion that takes place between students and teacher during this lesson is critical and should show a deeper understanding.

Reflect in Journals

What worked? What didn't work? What would they change? I used a rubric to assess each student's learning through their journals and anecdotal notes (see Internet Resources). Amazingly, the students totally forgot about the competition piece of the lesson—they were so excited about their successes and finally "getting it." One student, when writing about the changes to his group's design said, "I saw another group's plane fly real far and realized theirs looked way different than ours. When we did it again, we thought about what theirs looked like when we designed ours." Another said, "I can't always throw it super hard, you have to throw it soft sometimes to make it catch the air under its wings...."

This type of activity allows students to build their knowledge and gain that deeper understanding of concepts. There are many potential topics for design challenges; see Figure 7.3 for a list. We need to give students the opportunity to experience science before we start explaining science. Students are developing science skills, connecting to the real world, and developing conceptual knowledge all at the same time. Through engineering and challenging students, we are able to excite and build content knowledge in our students that will last not only until the next test, but for a lifetime ahead of them.

Internet Resources

Design Challenges

> *www.teachertube.com/viewVideo.php?video_id=153857&title=Design_Challenges&ref=Gerlacj*

Dryden Flight Research Center

> *www.nasa.gov/centers/dryden/history/HistoricAircraft/X-Planes/index.html*

Engineering Design Challenges

> *http://edc.nasa.gov*

FIGURE 7.3.

Topics to consider for design challenges

- Creating better levies: Coastal erosion
- Building a better lunar lander
- Creating the best reentry vehicle from space
- Designing the best shipping container
- Creating the best cooler to keep drinks cold in the summer
- Designing an animal to live in a specific environment
- Creating the perfect planter

PART TWO

Content Area Activities

CHAPTER 8

Repairing Femoral Fractures
A Model Lesson in Biomaterial Science

By Jarred Sakakeeny ///

Biomaterial science is a rapidly growing field that has scientists and doctors searching for new ways to repair the body. A merger between medicine and engineering, biomaterials can be complex subject matter, and it can certainly capture the minds of middle school students. In the lesson described here, seventh graders generally learn concepts such as systems in living things, properties of matter, and bioengineering technologies as they relate to adaptive and assistive technologies.

For this activity, two classes of seventh-grade students were given the task of repairing a 15-year-old boy's broken femur. Groups of students were given replica legs with identical femoral (thigh bone) fractures. Each group had to design bone plates, surgically implant their design and fasten it to the broken bone, and then suture the leg back up. The goals of this design project were to teach the engineering design process (see Figure 8.1, p. 52) and to introduce biomedical engineering with emphasis on surgical implantation and biomaterials. This lesson was used to complement curriculum on systems of the human body. The

Informational Packet can be downloaded for free and should be passed out to students, as it offers guided instruction, examples, and rubrics for grading (see Resources).

Students were first presented with the history and current applications of biomaterial science. Informational PowerPoint slides about this topic are posted on the Center for Engineering Educational Outreach at Tufts University website (see

Internet Resources). This website is free and gives interesting examples of how to better introduce engineering into different grade levels. Example biomaterials and devices were illustrated in the slides by elaborating on product evolution, reasoning behind material selection, and expectations for each device. Common biomaterials and devices such as contact lenses, pacemakers, and tendon grafts were discussed throughout the slide-show presentation, and students were engaged by simple questions such as, "What material do you think this is made of? Why would engineers use this material versus some other

material? How long should this device last?" and so on. The presentation can take as little as 20 minutes, or fill an entire class period depending on how in-depth the teacher chooses to go. Time should be allotted for laying out the goals and expectations of this project, along with emphasis on the eight steps of the engineering process (see Figure 8.1).

Keeping with the engineering design process, the problem was presented to students as follows "A 15-year-old boy has broken his femur—your job is to design a device/implant to fix/secure the bone and devise a surgical technique to implant your device." A helpful x-ray image of the type of fracture students would be attempting to fix (retrieved from an internet image search of "broken femur x-ray") was provided to allow students to visualize the severity of the fracture and the way in which they would have to fix it. Students were assigned to groups of four or five, and each group then brainstormed variables that they felt were important to research. We employed guided instruction during the brainstorming process to ensure that each group had an adequate and mostly equal starting point for their research. Each student was required to document his or her work through the packets, which were distributed at the beginning of the project.

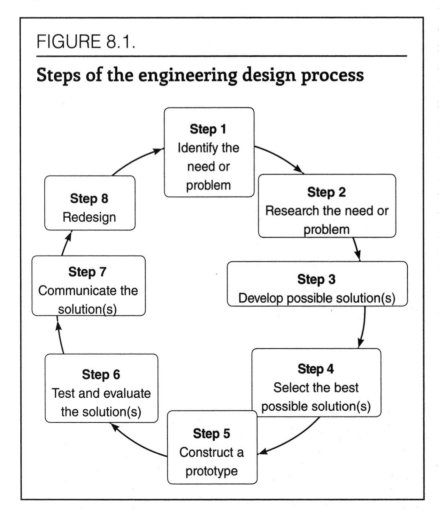

FIGURE 8.1.

Steps of the engineering design process

Step 1 Identify the need or problem

Step 2 Research the need or problem

Step 3 Develop possible solution(s)

Step 4 Select the best possible solution(s)

Step 5 Construct a prototype

Step 6 Test and evaluate the solution(s)

Step 7 Communicate the solution(s)

Step 8 Redesign

Construction of a Fake Leg

The construction of the leg can be messy, but quite safe and simple. Materials include foam padding, also called *underlayment* (typically laid down under hardwood flooring to prevent squeaking), some canned expandable foam insulation (the brand name is Great Stuff), and a 1 in. diameter wooden dowel. All materials can be purchased at a hardware store at a minimal cost. The foam padding represents the skin, the expandable foam represents the muscle tissue, and the dowel represents the broken bone. For five classes of 30 students, you will need 35 12 in. long dowels (sold in larger lengths to be cut down), 16 cans of Great Stuff, and one roll of the underlayment. The whole project should cost less than $130. The teacher should prepare the legs ahead of time. Instructions for assembling the artificial legs are as follows (see Figure 8.2):

1. Cut a square of the foam (roughly 14 inches) and spray out a layer of Great Stuff on top.

2. It is worth noting that Great Stuff will not come off of clothing no matter how hard you try, so wear clothes that you don't care about. A second layer may need to be applied after the first layer has dried. Wait 15 minutes for the foam to fully expand, but not fully harden, then rest the broken wooden dowel (two 6 in. pieces) on top.

3. Roll the foam so that the broken dowel is running along the length of the rolled up foam. Place a strip of duct tape over the seam to hold it together while the foam fully dries.

Legs should be prepared in batches of 5–10 at a time. Each batch will take about 30 minutes from start to finish with a lot of that in drying time. Creating legs for five classes can be done in less than three hours. Additional characteristics of the human leg may be added to make the leg more realistic. Veins and arteries can be added by purchasing flexible tubing. Blood can be added in between the muscle and skin layers by adding ketchup on top of the foam, laying down plastic wrap on top of the ketchup, and then adding Great Stuff. There are infinite possibilities that can be added, depending on the focus of the project or the preference of students. However, for every addition, there will be more complications and

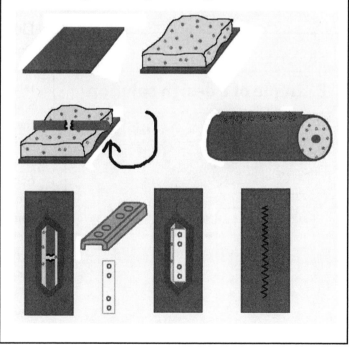

FIGURE 8.2.

The process of constructing and mending the replica leg

a higher cost. This project can also be done on a much smaller scale by replacing the expandable foam with another material such as foam sponge or another elastic durable product. Changing the femur bone with a smaller bone such as the humerus (arm bone) would also allow you to scale back the project a bit and get more for your money.

Researching the Problem

Using the internet, students researched everything from "How to fix a broken femur" to surgical tools and techniques in books and journals. An internet image search of "femur fracture" yields great images for students to visualize possible solutions. Students should be supplied with computers (with internet access), books, and journals (biomedical journals, *Popular Science*, and so on) to help complete this step. All work was documented and cited to fully illustrate the importance of research. Two class

periods are adequate for students to conduct research. The resource packet guides students through researching the right topics, but the teacher should urge them to discover techniques that real doctors currently use to re-align and fix severely fractured femurs.

The research step is the most important step for students to form teambuilding tactics and to fully develop a coherent strategy. This is the stage when team leaders often take charge. Because of the shared responsibilities throughout the project, it is important to be aware of how students are placed in each group. A mixture of students with various strengths and weaknesses should be paired together to promote this natural emergence of group leaders. We also find that the quieter students become subdued when other group members take charge. It is important for the teacher to stress the idea of teamwork and to delegate work early on to promote team cohesiveness.

Develop Possible Solutions

Once adequate research had been conducted, students were asked to individually create two separate design solutions for homework (Figure 8.3). Specific directions and examples are addressed in the packet, but designs had to address the problem statement, provide dimensions and other relevant details, and include a brief description of how the device/implant would work to fix the broken bone. For the following class, students compared their designs with those of their group members and evaluated each other's work. Students were asked to constructively criticize their designs, listing pros and cons for each design in order to reach conclusions about how they could develop their final design. Each group's final design was plotted using positive aspects of each of the individual designs. Prototypes would

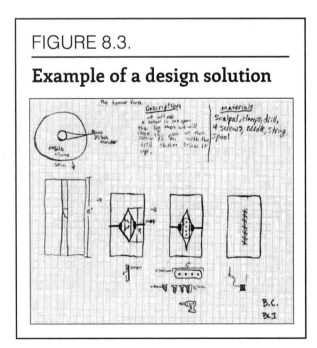

FIGURE 8.3.

Example of a design solution

specifically follow the final design, which made it critical that each design contained dimensions, materials, tools needed, and a description of how it would work.

Prototyping and Implantation

Students were eager to begin work on prototyping their designs. They brought in recyclable materials such as corrugated cardboard, plastic, soda cans, foam rubber, and duct tape. The class recognized the fact that they would never put those materials into a human body, but that they were not expected to be using titanium for the purposes of their bone plates in the classroom. Prototyping took place over two class periods (45-minutes each class, minus set-up and clean-up time), following the design specifications that each group had finalized. Each group was prepared for surgical implantation by the end of the second day of prototyping. Homework assigned for each group required students to explicitly write down each step of their surgical procedure starting with "Make an incision [blank] inches long" and ending with "Suture the skin of the leg to close the incision."

Groups were given tools for cutting skin, muscle removal, fastening their devices into place, and stitching. Box cutters were borrowed from the tech-ed shop, along with screwdrivers and hammers. Plastic needles were purchased along with string for stitching the legs. Surgeries took place over two days, with all groups successfully implanting their devices and all group members taking part in the surgery.

Safety is a major concern during this step of the project. Leather gloves should be used to protect hands during the cutting of "muscle." (One set of gloves per group is enough.) Because students may find the muscle tissue hard to cut through, they should be given safety lessons

before using the box cutters. Extra adult supervision is helpful at this step of the project to watch over groups as they begin cutting.

Students will operate on the side of the leg directly opposite the duct tape. Teachers should remind students of the following rules:

- One person cuts at a time.

- Never reach for or grab the box cutter from someone.

- Keep fingers away from the cutting edge.

- With the object on the desk, stand in front, hold the object firmly above the line of cutting with your other hand, and cut in a direction toward the body.

It is up to your best judgment to decide who should be using the cutting devices and what level of training is necessary. I was fortunate enough to have classes familiar with cutting tools and safety because of required classes in tech-ed shop. If you are unsure how to operate the tools, ask for help prior to teaching this lesson.

Once inside the leg, the groups can then implant their devices and secure the bone (see Figure 8.4, p. 56). The final step is to stitch the leg back up. For these sutures, thick plastic needles can be used instead of sharp metal needles. These plastic needles are intended for yarn work and can be purchased at any arts and crafts store.

Communicating Solutions

The final requirement of this project involves each group giving 10–15 minute PowerPoint presentations to the class to explain the steps that they took to solve this problem. Again, the requirements, instructions, and rubric have been provided in the supplemental packet and can be referenced at any time by students. Students revisited their work and described their thoughts, problems, solutions, and what they would change throughout the project. Each group member was required to speak and explain the reasoning behind their decisions.

Students' explanations, analysis of their problems, and the logic behind a redesign allowed for an assessment of their knowledge gained from this project.

FIGURE 8.4.

Example of a surgical procedure in progress

A. Students make their primary incision to expose the muscle tissue. B. The broken femur is located and evaluated. C. Students drill the prototype into place to secure the bone. D. Stitching the leg back up.

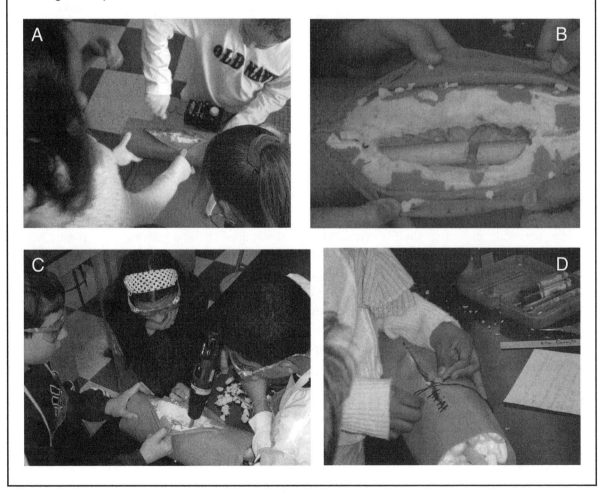

Assessment

Assessing students in this project was done in several ways. Students' work was tracked throughout the project using the previously described packet. The packet outlines expectations and instructions for each step of the design process and asks students to explain their methods. The packet can be used to differentiate members within a group to better understand each student's comprehension of the subject matter. The packet also includes sample design sketches pertaining to a different activity conducted in another class. The sample sketches gave students a better understanding of our expectations for their design sketches, but did it in a way that did not unduly influence them to produce a preconceived possible design.

A postassessment of the project was also administered in the form of an informal quiz (available online; see Resources). This quiz was intended to reveal what students had learned about

- the engineering design process,
- the human body on a macro and micro scale,
- different biomaterials and their applications, and
- how they would have changed any aspect of the project to make it better for them.

By removing the long-established written exam as a form of assessing and grading students, some of the more visual and hands-on learners were able to showcase their skills and flourish. Students were no longer intimidated by the idea of one correct answer and instead allowed their creativity to flow into what they perceived to be the best answer.

Engineering Advantage

This project was successful in its incorporation of the engineering design process into the classroom because we were able to integrate several aspects of the seventh-grade science curriculum into one activity without disrupting the flow of learning. This activity promoted scientific thought by allowing students to follow their scientific intuition instead of using an explicit set of predetermined directions. Standards set forth by the American Association for the Advancement of Science are satisfied by this activity in the form of:

- Abilities necessary to do scientific inquiry
- Ability to design a solution or product
- Ability to communicate scientific procedures and explanations
- Understanding that a substance has characteristic properties
- Understanding the structure and function of body systems and their interrelationships
- Understanding that technological solutions have intended benefits and unintended consequences

This project was a successful merger of science and engineering into one classroom activity. It is a fun activity to conduct because it is simple to modify and customize to fit each class and optimize knowledge transfer. Expectations were set high on this project, and students were well aware of the difficult subject matter that they would be covering. Despite the high expectations, they were willing to put forth an effort that led to the success of their projects.

Acknowledgments

The author would like to acknowledge the help and guidance of Katie Cargil, Meredith Knight, and Brian Gravel.

Internet Resources

Center for Engineering Educational Outreach at Tufts University

www.ceeo.tufts.edu

Instructional packet

http://130.64.87.22/cool/sakakeeny/Biomaterial_ steps.doc

Introductory slide show presentation

http://130.64.87.22/cool/sakakeeny/Biomaterial_ Science.ppt

Post assessment quiz

http://130.64.87.22/cool/sakakeeny/assessment. doc

Guided surgical technique

http://130.64.87.22/cool/sakakeeny/surgical_ technique.doc

CHAPTER 9

Get a Grip!
A Middle School Engineering Challenge

By Suzanne A. Olds, Deborah A. Harrell, and Michael E. Valente ///////////////////////////////////

Investigating the field of engineering offers the opportunity for interdisciplinary, hands-on, inquiry-based units that integrate real-world applications; yet, many K–12 students are not exposed to engineering until they enter college. Get a Grip! is a problem-based unit that places middle school students in the role of engineers who are challenged to design and construct prosthetic arms for amputees in a war-torn country. The students use common materials to build arms that accomplish tasks requiring fine motor control and strength. A critical component of the unit is its ability to demonstrate to middle school students that strong, interdisciplinary knowledge is required to solve engineering problems. As such, it is a practical and efficient mode of interdisciplinary instruction meeting state and national standards in science, math, reading, and social studies.

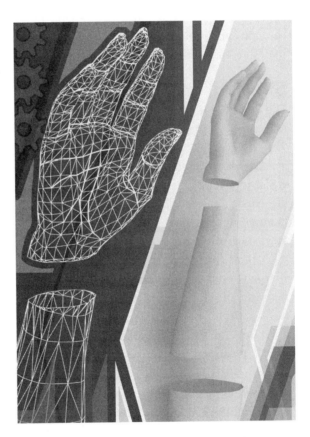

This activity, the result of a partnership among university faculty, K–12 teachers, the Center for International Rehabilitation, and university engineering students, seeks to:

- inform middle school students about engineering as a career—what engineers do and the impact they have on society,

- engage middle school students in the engineering design process, and

- encourage middle school students to draw on previously learned science concepts to accomplish a real-world engineering task.

Get a Grip! is a variable-length unit that challenges middle school student teams (groups of four or five) to design and construct a prosthetic arm from common materials for a 12-year-old Afghan girl who needs to eat and carry water from a nearby river to her home. Limiting the supplies to those that are readily available in that country constrains the students and reduces the materials cost of the unit.

Structure of the Unit

The Get a Grip! unit is composed of eight lessons that support the Grand Challenge (see Figure 9.1). This curriculum is available online at *www. middleschoolengineers.com*. The cost to participate is $50, which includes training and support to use the curricular materials, access to the teacher's manual (lesson plans, student handouts, teacher notes, answer keys, extension activities), plus the videos referenced in this article. This access charge will be used to sustain the support, training, and development of the module—and not for any profit. For the 2006–2007 academic year only, the NSF grant will cover the access charge for all participants. The training is online and can be taken at the user's convenience. Tools that can also be accessed from the site including a document repository, a discussion forum, and an e-mail list. The cost does not include the materials used to build the arms or the Pinto's Hope books. The materials to build the arms can be purchased and gathered for about $10–15 per box, with a much smaller replenishment cost. The *Pinto's Hope* books can be purchased from *www.iuniverse.com* for about $8 each, depending on the quantity ordered. Each lesson may be adjusted in length based on the content goals of the teacher. All lessons follow the Legacy Cycle framework, a format that incorporates findings from educational research on how people best learn (see Figure 9.2, p. 62). Research and theory behind this method of learning may be found in *How People Learn* (Bransford, Brown, and Cocking 1999). All lessons also include extension activities that enable the teacher and students to further explore some of the topics addressed. If all lessons are completed in full, the unit will take about 30 hours of classroom time (about seven weeks if done entirely within a science classroom). However, some teachers have trimmed this unit down to three or four weeks, depending on their needs. The unit can be reduced in length by offering some activities concurrently in other subjects. This project has been implemented by middle school teachers who are teaching over 140 students at any given time. They have found it helpful to keep the materials for the arms in one clear plastic box for each team. Each team can store their materials, sketches, and notes in the container. Teachers have also found it helpful to assign a role to each group member (facilitator, time keeper, recorder, spokesperson, and so on) and to distribute team evaluation sheets periodically to keep each group focused.

Before beginning the unit, students may gain an appreciation for the culture and plight of the amputee by reading the award-winning novel *Pinto's Hope* by module co-developer Deborah A. Harrell. While helpful, the book is not an essential part of the unit. The book is 64 pages and has questions within it for discussion. Many teachers have assigned this as outside reading or through the language arts class. *Pinto's Hope* is a story of a young boy's recovery from a land mine accident and his adaptation to his new prosthetic leg. The literature component to the unit offers cross-curricular enhancement lessons that sup-

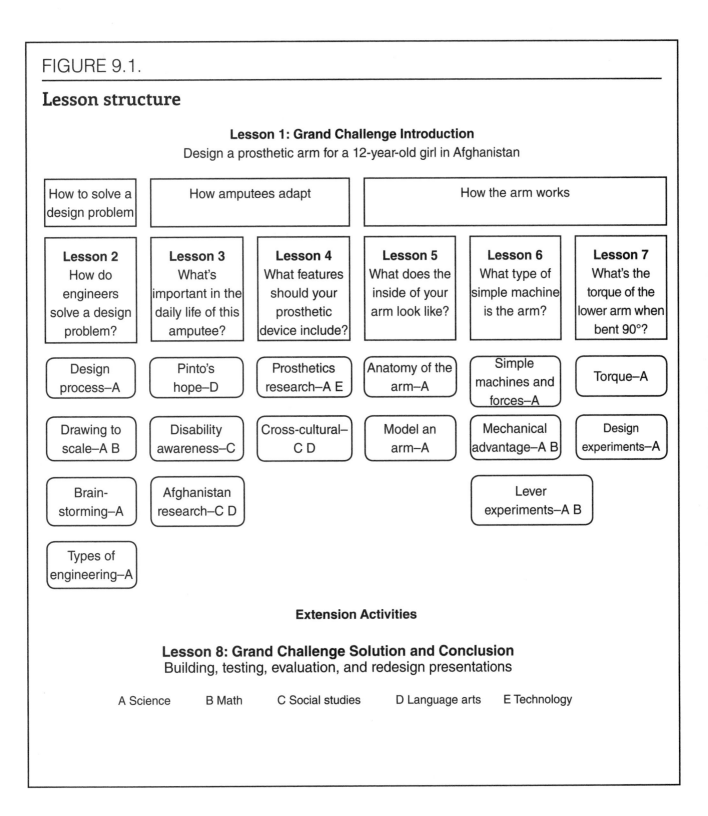

FIGURE 9.1.

Lesson structure

Lesson 1: Grand Challenge Introduction
Design a prosthetic arm for a 12-year-old girl in Afghanistan

How to solve a design problem	How amputees adapt	How the arm works

Lesson 2
How do engineers solve a design problem?

Lesson 3
What's important in the daily life of this amputee?

Lesson 4
What features should your prosthetic device include?

Lesson 5
What does the inside of your arm look like?

Lesson 6
What type of simple machine is the arm?

Lesson 7
What's the torque of the lower arm when bent 90°?

Design process–A

Pinto's hope–D

Prosthetics research–A E

Anatomy of the arm–A

Simple machines and forces–A

Torque–A

Drawing to scale–A B

Disability awareness–C

Cross-cultural–C D

Model an arm–A

Mechanical advantage–A B

Design experiments–A

Brain-storming–A

Afghanistan research–C D

Lever experiments–A B

Types of engineering–A

Extension Activities

Lesson 8: Grand Challenge Solution and Conclusion
Building, testing, evaluation, and redesign presentations

A Science   B Math   C Social studies   D Language arts   E Technology port many of the Standards for English Language Arts (SELA).

On the first day of the unit, students are introduced to the Grand Challenge. The Grand Challenge, like the other challenges that initiate each lesson, is a question that engages students, and piques their curiosity. The Grand Challenge frames the unit and requires students to bring to bear their current knowledge and preconceptions about the topic. After viewing a downloadable five-minute video of the Grand Challenge, students generate ideas by writing responses to the following questions in their journals:

- What do you know about this problem?

- What ideas do you have for the prosthetic arm?

- What information do you need to solve this problem?

FIGURE 9.2.

Legacy Cycle framework

Challenge	A question that causes students to wonder about the topic and become engaged with it. The question frames the unit or lesson and requires students to bring to bear their current knowledge and preconceptions about the topic.
Generate ideas	A whole-class activity that causes students to display and compile their current knowledge/ideas/perceptions. Implementation of this step is often done in the form of questions: What things would you need to know to answer this question? What additional information would you like to have to help you answer this question?
Multiple perspectives	Outside resources that provide information related to the topic of the challenge. These tend to "point students in the right direction" for further inquiry.
Research and revise	Additional information that students receive/seek. This may be in the form of inquiry-based experiments, lectures, readings, websites, and so on. Students revise their original ideas based on new information.
Test your mettle	A set of activities in which students engage to help them explore their depth of knowledge. The goal is to create formative assessment situations that help them evaluate what they do not know so that they may return to the research-and-revise step again to learn more.
Go public	Final conclusion(s) that students display.

The technique is similar to the KWHL technique for helping students activate prior knowledge, gather information, and think through a problem. (Students write what they **K**now about the problem, what they **W**ant to know about the problem, **H**ow they will find information to solve the problem, and what they have **L**earned.)

The students then obtain multiple perspectives by sharing their ideas with the class. We then categorize their responses, grouping them as much as possible into the subsequent challenges that will be addressed in the lesson:

- How do engineers solve a design problem?

- What is important in the daily life of this amputee?

- What design features are important to include in your prosthesis?

- What does the inside of your arm look like?

- What type of simple machine is the arm?

- What is the torque of your lower arm when it is bent 90°?

The second lesson has the class investigating how engineers solve design problems. To help generate ideas, the class considers how they will approach the problem of designing a prosthetic arm for a 12-year old girl, Laila, from Afghanistan. In this lesson, multiple perspectives take two forms—students hear their classmates' ideas and they watch a short video of engineers at work. The teacher plays the video one or two times, after which students revise their initial ideas about how an engineer solves a problem. Students gain additional insight into how engineers solve a problem in the research-and-revise activities of this lesson. In this second lesson, research-and-revise takes the form of team-based activities (brainstorming, drawing to scale and from multiple perspectives)

that mirror the work of real-world engineers in search of a solution to a design challenge. In other lessons, research-and-revise takes the form of inquiry-based experiments, minilectures, readings, and internet research. This lesson, like all others, concludes with a formative assessment activity (usually a quiz or journal activity). Finally, students are asked to go public with the knowledge they have gained. This can be done in various ways—journal writing, drawing a cartoon, producing a brochure, creating a model, or giving a presentation. In this particular lesson, it is done through a journal activity titled "How do the activities in this lesson relate to the Grand Challenge?"

In lessons 3 and 4, students generate ideas about features their prostheses should include. To advance their initial ideas, students engage in a disability awareness activity and investigate cultural issues and prosthetic design. In the disability awareness activity, students discover some challenges amputees face. They try to tie their shoes, carry water, and make a jelly sandwich with one arm in a fist and the other arm functioning as usual. Another research and revise activity is researching the culture, geography, and demographics of Afghanistan via a guided internet search. Students then engage in a cross-cultural comparison, where they compare the needs of an amputee in Borneo to those of Jessie Sullivan, the world's first "Bionic Man." The final research and revise activity is an investigation of different types of prostheses and their uses.

All activities together prompt students to think about the user of their prosthesis and any special needs she may have due to culture, geography, or demographics. Assessments in the "test your mettle" section include a quiz and journal responses to: "How is life in your country different than life in Afghanistan? What do you think are the most important differences between an artificial limb you might use and one the amputee would use? In the "Go Public" section, students respond in their journal to the following questions: Besides eating and carrying water, what other daily tasks might the Afghan girl need to accomplish? What qualities or features might you include in your prosthesis to make these tasks easier for her? There are five extension activities related to these lessons. One example is to have students write their senator, urging the United States to sign the Ottawa Mine Ban Treaty. Another extension activity is to have students carry out their daily school activities for one entire school day using only one arm.

Lesson 5 transitions students to examining how a human arm operates. After students attempt to draw all the components in their arms (generate ideas), they are asked to bend their arm and discuss what makes the arm bend (multiple perspectives). In the research-and-revise section, they explore the anatomy of the arm via PowerPoint slides and an activity where they model a functional arm with cardboard and balloons. In the test-your-mettle section, students take an anatomy quiz and either journal or discuss the following questions: "Why is it necessary for doctors and biomedical engineers to learn about the anatomy of the human body before they can treat patients or create prosthetic devices for patients? What arm components is Laila missing (completely or partially) due to her amputation? What are some features you might include in your arm to compensate for the missing parts? If Laila's arm was amputated across the humerus instead, would you design your arm differently? If so, how?" In the go-public section, students revise their initial arm sketches and also create a model of a functioning body part.

Lessons 6 and 7 bring the students' attention to the biomechanics of the arm, as they engage in inquiry-based simple machines activities. Students are challenged to determine what type of simple machine the arm is, what mechanical advantage that offers, and how much effort it takes to keep their arm bent at a 90° angle. To address these questions, they engage in a Rube Goldberg activity and design their own experiments to determine the relationship between:

1. load position and effort;

2. load position and mechanical advantage; and

3. force, distance, and torque.

Students then relate the knowledge they gained in these lessons to the Grand Challenge by addressing questions such as: "Should the torque of your prosthetic arm be minimized or maximized? How could you achieve that?"

Students conclude the unit (Lesson 8) by addressing the Grand Challenge, which requires them to synthesize and apply the information they gained in the previous lessons. In this final lesson, the students design and build a prosthetic arm that meets the requirements of the Afghan girl. There is a suggested materials list included in the online packet; however, it does not need to be followed exactly and many items can be found in the classroom or by having students bring them in. Common items used by students in their final designs include cotton rags, pantyhose, plastic bottles, kindling wood, metal hooks, PVC pipe, and a lot of duct tape. Supplies can be purchased for about $10–15 per student team, and restocking is on the order of $5 per box.

Students should be given at least four class sessions to complete this phase of the design, although giving them six or seven sessions is better. Students work in teams of four or five. Once they build a prototype, they test it performing specific tasks (carrying water a distance of 10 m and setting it on a table 1 m high; and, in less than a minute, lifting three olives to their mouths one at a time without piercing their skin). Then they evaluate the arm on a testing rubric that is provided (or can be designed by them). The testing rubric includes the design requirements the students came up with (such as comfort, cost, adjustability, and ability to do the tasks). Eight design requirements are provided and the teacher can select those appropriate to each class. When their final design is complete (usually determined by time constraints rather than feeling they are "done"), their arms are tested and evaluated by a different group. The other group will also disassemble the arm, check the cost of the arm, and inventory the supply box.

Students "go public" with their final designs by making an oral presentation to the class. In this presentation, they cite the strengths and weaknesses of the arm and discuss what one additional material they would have liked to have used and how they would have used it. They also produce a brochure that advertises their artificial limb to a specific target audience (potential users of the limb or organizations that are interested in purchasing the limb). Extension activities include commercials, skits, and government proposals.

Meeting Standards

Get a Grip! A Middle School Engineering Challenge was developed in alignment with both the National Science Education Standards (NRC 1996) and the *Benchmarks for Science Literacy* (AAAS 1993). Each of the eight core lessons correlates with one or more of the standards. Since most states base their standards on these national guidelines, all of the lessons meet, and in some cases exceed, state and other local standards.

Additionally, most lessons address objectives for language arts, reading, mathematics and social studies, supporting National Science Education Standards Program Standard B, which states, "The program of study in science for all students should be developmentally appropriate, interesting, relevant to students' lives; emphasize understanding through inquiry; and be connected with other school subjects." The guiding principal of Get a Grip! is summed up in the AAAS (1993) Benchmark 3a (The Nature of Technology—Design and Systems): "Perhaps the best way to become familiar with the nature of engineering and design is to do some" (p. 48).

Assessments and Conclusions

It is only appropriate to ask if students "got a grip" because of this project—that is, were the students able to demonstrate increased understanding of the engineering design process and an improved ability to apply it? Did they know more about engineering as a career? Did their understanding of basic science concepts improve? Did students gain awareness that technological design involves many other factors in addition to scientific issues? Yes, to all of the above! Several assessment tools were used to measure the effectiveness of each lesson. Pre- and postproject homework assignments and surveys helped us assess students' science, math, and engineering content knowledge as well as their views on these topics. The results indicate that participation in the project increased their understanding and interest in engineering, their enjoyment of science, and their simple machines content knowledge.

Because Get a Grip! explores the unusual topic of amputees, it engages the students like no other. The 1,000+ students who have tested the unit all report a high "fun" factor and a high "learn" factor. Most of the students can remember the design steps well after the unit's completion. Teachers as well are excited by its interdisciplinary nature, incredible efficiency, and links to current real-world problems.

Acknowledgment

Get a Grip! was initiated in September 2001 with funding from the National Science Foundation. This work was supported primarily by the Engineering Research Centers Program of the National Science Foundation under Award Number EEC-9876363.

References

American Association for the Advancement of Science (AAAS). 1993. *Benchmarks for science literacy*. New York: Oxford University Press.

Bransford, J. D., A. I. Brown, and R. Cocking, eds. 1999. *How people learn: Mind, experience, and school*. Washington, DC: National Research Council.

Klein, S. S., and A. Harris. 2005. *Electrocardiogram mosaic: Vanderbilt instruction on biomedical engineering for secondary science (VIBES)*. Nashville, TN: Vanderbilt University.

National Council of Teachers of English (NCTE) and International Reading Association (IRA). 1996. *Standards for the English language arts*. Urbana, IL: International Reading Association and the National Council of Teachers of English.

National Research Council (NRC). 1996. *National science education standards*. Washington, DC: National Academies Press.

Internet Resource

Center for International Rehabilitation
www.cirnetwork.org

CHAPTER 10

A Partnership for Problem-Based Learning

Challenging Students to Consider Open-Ended Problems Involving Gene Therapy

By Amanda Lockhart and Joseph Le Doux ///

So many demands are placed on high school teachers today. We are expected to engage students in authentic hands-on, inquiry-based learning that hooks them on the excitement of cutting-edge science fields, develops their higher-order thinking skills, exposes them to modern lab techniques, arms them with an understanding of the science that affects their lives and should inform their decisions as adults, and covers state and national curriculum standards. Simple? Not really. But the right kind of professional development can help teachers meet these challenges.

For two summers, I took part in the Research Experience for Teachers (RET). Through this RET professional development partnership, I attained greater in-depth knowledge of molecular biology and virology, became more familiar with cutting-edge lab techniques that teach students important skills for college, and brought my summer research experience back to the classroom through a problem-based learning (PBL) gene therapy module.

Background on RET Partnership

When most teachers hear the phrase "professional development," skepticism clouds their faces. But research-based professional development can provide interesting experiences and valuable content knowledge. Though many teachers do

participate in research, the National Science Education Standards suggest that more should. Professional Development Standard A states that a science teacher's learning experiences must "involve teachers in actively investigating phenomena that can be studied scientifically, interpreting results, and making sense of findings consistent with currently accepted scientific understanding" (NRC 1996, p. 59).

The RET program, funded by the National Science Foundation, helps "facilitate professional development of K–12 teachers and community college faculty through strengthened partnerships between institutions of higher education and local school districts; and encourages researchers to build mutually rewarding partnerships with teachers" (NSF). The long-term goals of our partnership established through RET are to stimulate the development and use of educational techniques and materials that will inspire high school students to consider careers in science and engineering and to increase communication between high school and college educators.

For 10 months of the year I am a high school biology teacher. For the other two months I am a research scientist, at least I have been for the past two summers when I have worked in the Le Doux Lab at Georgia Tech as part of the RET program. Research in the Le Doux Lab revolves around gene therapy, focusing on retrovirus and lentivirus-mediated gene delivery to cells and the effect of genetic modification on the fate of embryonic and adult stem cells. My participation in the RET program has enabled me to help establish a three-member partnership that includes Joseph Le Doux, an associate professor of biomedical engineering, Cindy Jung, a biomedical engineering graduate student, and me.

The first objective of our partnership was to develop curriculum materials to motivate secondary school students to learn the basic concepts of gene therapy, understand what this technology can and cannot do, and consider what its effect could ultimately be on their health.

Gene therapy is a rapidly emerging technology with the potential to have a profound impact on medicine and society. It provides an excellent example of how today's scientists and engineers must integrate principles from several disciplines (physics, chemistry, biology, and engineering) to solve problems. In addition, the excitement generated by gene therapy may inspire some otherwise disinterested students to consider a career in science or engineering.

My first order of business was to get up to speed on gene therapy and molecular biology techniques. I learned cell culturing techniques and the planning, calculations, and techniques involved in several experimental assays. During my first summer (2003), I conducted experiments that helped show that the half-life of lentiviruses was longer than a related family of viruses called oncogenic retroviruses. The following year (2004), my investigations focused on developing methods to modify the properties of viruses in novel ways that are expected to have future use in engineering viruses to target and genetically modify specific cell types for the treatment of diseases or to help create artificial organs.

These research experiences forced me to relearn how to approach experimental design when I know little about the problem. I now view experimental design as a creative process rather than the step-by-step method often taught in classrooms. I realized that I need to help my students learn about the scientific process rather than teach them the "scientific method." It became clear to me that becoming an expert on an intriguing topic is the most effective way to learn important concepts and to be able to generalize those concepts.

FIGURE 10.1.

"What Genes Are You Wearing?" PBL module

Hook: Attention getter designed to activate prior knowledge and inspire further study. Students listen to a segment from a National Public Radio interview with a young woman living with cystic fibrosis and reflect on the following questions:

- How would you feel if this were you?
- How would you feel if this were your child?
- What are the risks and benefits of a lung transplant?

Challenge: Gives students ownership of their learning by guiding them to learn about gene therapy so they can make an informed decision. The scenario is as follows:

You are a 16-year-old who suffers from [insert disease here]. Recently your condition has been worsening and you have been spending an enormous amount of time in and out of the doctor's office and hospital. You are tired of being sick and just want the opportunity to be a teenager. Your primary physician has recently suggested that you undergo gene therapy.

You have no idea what gene therapy is or how it works. Your doctor keeps stating that you are lucky because there is a new study starting within the month at your hospital to test a new form of gene therapy for your disease. The gene therapy is very experimental and still in the very early stages, and the outcome of the treatment is uncertain. You have a week to decide if you want to become a part of this study, but before you decide if you want to join the study you must do some research on the topic.

Generate ideas: Students join groups and work to solve the problem based on their preexisting knowledge by considering three questions.

- What do you know?
- What questions do you have?
- What questions will require further research?

Meet the experts: Experts discuss their point of view about how to approach the challenge. Students review two video clips that discuss important points to consider about gene therapy and the risks versus benefits of gene therapy.

Research and revise: Students conduct research to help them meet the challenge. Students complete questions about diseases and gene therapy. Students are provided with the instructions:

- List questions you have.
- Answer all questions.
- Share and compare with other group members.

Test your mettle: Students engage in a mini-challenge to evaluate what they have learned. Students are provided with a list of diseases and medical conditions such as cystic fibrosis, muscular dystrophy, Alzheimer's disease, and others and asked to determine which would be appropriate for treatment by gene therapy.

- Students compare their answers to those of the experts and discuss why their answers may vary.
- Students do further research as necessary.

Go public: Students present and defend their solution in public:

- Individually, students decide with their parents or guardians if they will join the clinical trial.
- Groups reach a consensus in class.
- Groups defend their decision before the class.

Problem-Based Learning in High School

As I became more comfortable with the research, I began to think about how I could effectively bring what I had learned back to my classroom. Shortly after I arrived, Le Doux shared his vision for implementing PBL in the high school classroom. His department had been successful in modifying PBL—a technique commonly used in medical schools, for use in engineering and science courses. They found that PBL, by placing their students in situations that mimicked those faced by practitioners in the field—was an extremely effective way to

---

FIGURE 10.2.

Gauging student learning

To gauge student learning, students received a pre- and posttest consisting of the same five questions:

- Do you know what a genetic disease is? Give a definition.
- Give an example of a genetic disease.
- Do you know what gene therapy is? Give a definition.
- Name a disease that can potentially be treated using gene therapy.
- Do you know what genetic engineering is? Give a definition.

Students who answered "Yes" and could give a correct answer were put in the "Yes" category, while students who said "Yes" but could not give a correct definition or example were put in the "Not Sure" category. Students who answered "No" were put in the "No" category.

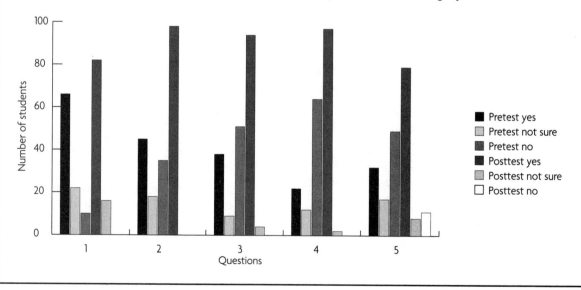

---

teach students how to think like biomedical engineers and scientists.

I had previously been intrigued by the idea of PBL, and partnering with Le Doux and Jung provided me the opportunity to try it out. I decided to use the "Star Legacy" inquiry-based learning method developed by The Iris Center at Vanderbilt University (IRIS). In this method, students are presented with a problem or challenge and are asked to outline an approach for addressing the challenge based on their current knowledge of the subject. Once they establish a plan, students learn from experts, who share the approaches they would have taken had they been asked to address the challenge. Armed with this additional insight, students improve their plan and proceed to implement it.

Later, students participate in exercises designed to determine how much they learned about the subject while tackling the challenge. Finally, they

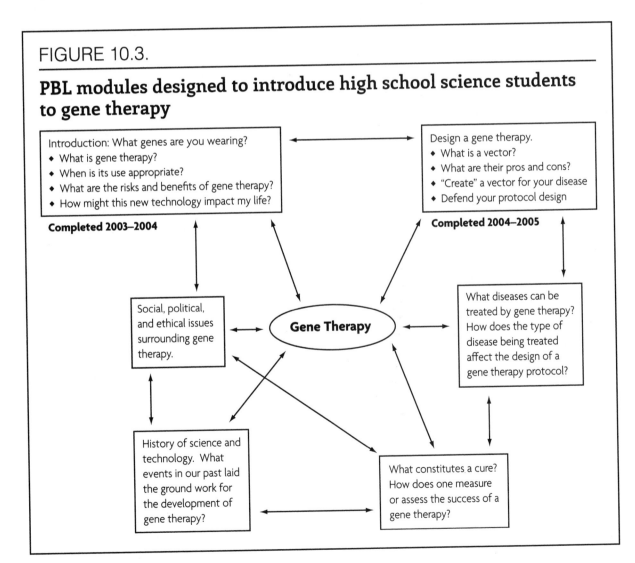

FIGURE 10.3.

PBL modules designed to introduce high school science students to gene therapy

Introduction: What genes are you wearing?
- What is gene therapy?
- When is its use appropriate?
- What are the risks and benefits of gene therapy?
- How might this new technology impact my life?

Completed 2003–2004

Design a gene therapy.
- What is a vector?
- What are their pros and cons?
- "Create" a vector for your disease
- Defend your protocol design

Completed 2004–2005

Social, political, and ethical issues surrounding gene therapy.

Gene Therapy

What diseases can be treated by gene therapy? How does the type of disease being treated affect the design of a gene therapy protocol?

History of science and technology. What events in our past laid the ground work for the development of gene therapy?

What constitutes a cure? How does one measure or assess the success of a gene therapy?

share their accomplishments with others inside and outside of the classroom. A key to the success of PBL is that the problem must be open ended and not strictly defined. This creates a situation much like what scientists and engineers face in the real world.

A Gene Therapy PBL Exercise

I created the "What Genes Are You Wearing?" PBL module (Figure 10.1, p. 69) in which students are faced with the following challenge: "You have a disease and have been given the opportunity to join a gene therapy clinical trial. You must decide whether or not to participate in the trial." The module began with a "hook" to excite students' interest—they listened to a previously recorded National Public Radio interview of a young cystic fibrosis patient to give them perspective on what it is like to be a teenager afflicted with an incurable disease (Rothenberg 2002).

Students then entered the "challenge" phase of the PBL problem. I presented them with a scenario in which each of them suffered from a serious genetic disease with no viable medical treatment. Each had been offered the opportunity to participate in a gene therapy clinical trial for their particular disease. Students then needed to decide whether or not to join the clinical trial and to back up their decision with facts about gene therapy.

To help students get started on their research, we had them "meet the experts" by watching a film in which Le Doux and another gene therapy expert, Peter Thule (Atlanta VA Medical Center and Emory University School of Medicine), discussed some of the key issues they would consider if deciding whether or not to join a clinical trial. My students then began to "research and revise" their decision, becoming gene therapy experts in the process.

After students had developed some expertise in gene therapy, I "tested their mettle" by asking them to comment on whether or not gene therapy was justified for a number of diseases and conditions, including baldness, familial hypercholesterolemia, cancer, and heart disease. To evaluate the soundness of their judgment, as a class we compared student decisions to the opinions of the experts.

Finally, students discussed their findings with their parents or guardians, further disseminating information about gene therapy to the public. Students reached a decision with their parents or guardians and then shared and justified their decision to their classmates in the "go public" portion of the PBL exercise. Each of these steps can be supplemented with alternative activities from various websites including the "Genetic Science Learning Series" at *http://gslc.genetics.utah.edu/ units/genetherapy*. For example, students can investigate current clinical trials at *http://clinicaltrials. gov*, where they will find regularly updated information about clinical research in gene therapy and other research in human volunteers. A valuable reference to the genetic disorders and gene therapy can be investigated at *http://ghr.nlm.nih.gov*.

Just the Beginning

It is important to note that PBL can be used with students of virtually any age and skill level because it does not require students to learn one specific set of facts. Rather, PBL encourages students to make explicit their thought processes as they work through problems, helps them improve their problem-solving skills, and by its very nature engages students at a level appropriate for their current age and degree of knowledge. I used the gene therapy PBL module in my general level classes with little modification and with great success.

Before implementing the PBL module in my classes, I gave the students a pretest (Figure 10.2, p. 70) that asked them basic questions, such as, "Do you know what gene therapy is?" If they answered "Yes," I asked them to give me a definition. Only 35 of 97 students knew what gene therapy was and could give an appropriate definition. After completing the PBL module, 93 of 97 students gave an excellent explanation of what gene therapy is and were able to name a disease for which gene therapy is currently being considered as a treatment.

I plan to develop and seek additional PBL modules and share them with my colleagues (Figure 10.3, p. 71). My partners and I have already completed a second PBL module designed to teach students about virus infection and virus structure, and we plan to post all of our PBL lesson plans, including the video of Le Doux and Thule, on the internet (see *www.bme.gatech.edu/groups/ledoux/* for a link to these materials).

Thanks to my RET experience, I have a better understanding of gene therapy and how gene therapy research is done in the university laboratory setting, and I have an enhanced understanding of laboratory techniques and experimental design, which I have already brought back to my classroom. In addition, I now have contacts that I can turn to for help with questions that will undoubtedly arise in a classroom where students are challenged to consider open-ended problems and plan how to move toward sound solutions.

References

National Research Council (NRC). 1996. *National science education standards.* Washington, DC: National Academies Press.

National Science Foundation (NSF). Research experience for teachers (RET): Supplements and sites. *www.nsf.gov/pubs/2003/nsf03554/nsf03554.htm.*

Rothenberg, L. 2002. *National Public Radio: All Things Considered.* My So-Called Lungs. August 5. *www.npr.org/programs/atc/features/2002/aug/socalledlungs.*

The IRIS Center (IRIS). What is a star legacy module? Star legacy modules. *http://iris.peabody.vanderbilt.edu/slm.html.*

CHAPTER 11

Plastics in Our Environment

A Jigsaw Learning Activity

By Elaine Hampton, Mary Ann Wallace, Kristan Keele, and Wen-Yee Lee ///////////////////////////

In this lesson, a ready-to-teach cooperative reading activity, students learn about the effects of plastics in our environment, specifically that certain petrochemicals act as artificial estrogens and impact hormonal activities. Much of the content in this lesson was synthesized from recent medical research about the impact of xenoestrogens and spun off from a curriculum project sponsored by the U.S. Army Research Office.

Lesson Content

The content for the lesson draws from the chemistry of petroleum products. Materials made from petroleum products surround us—medicines, lipstick, plastic bags, medical supplies, furniture, pesticides, and automobile parts, to name a few. Most plastic is made from petroleum products. Plastics and other petroleum products enrich our lives and have led to many valuable tools for society, including life-saving technologies such as medical tubing and artificial limbs.

However, artificial estrogens called xenoestrogens, associated with the production and use of plastics and other petroleum products, impact our health and our environment. Women and children are particularly vulnerable, as the by-products accumulate in breast tissue and affect fetal development. Most of the chemicals in plastics decompose so slowly that they are

basically permanent in the environment, or they decompose into other chemicals. Because plastics may affect our environment for a very long time, scientists need to explore their impact more thoroughly and search for alternative technologies. This lesson explores these effects in more depth.

Lesson Procedure

This is a cooperative-learning, reading, and discussion activity. Students read and discuss four essays that address the impact of synthetic chemicals on human health and animal reproductive organs, and the impact of plastic waste on the environment. The authors created the reading for this activity by reviewing recent and relevant medical research studies compiled by MEDLINE, an online medical research database. This information was then synthesized into four essays that are appropriate for middle school students.

The Jigsaw Approach

For this activity, students are placed into four "expert" groups. Each group is assigned a different article to read. While the articles contain technical terms that may seem daunting, by working cooperatively, students are able to master the vocabulary and the content. The members of the group can read the article silently, or the teacher can pair strong and weak readers and ask them to read the essay aloud to each other. After the groups have finished reading their articles, the group leader asks members to share what they think are the key points of the article. After the discussion, the members of each group separate, creating individual jigsaw pieces that then join together with pieces (students) from each of the other groups to create a jigsaw group. Students in these new jigsaw groups then share what they learned from their articles. (For more information about the jigsaw approach, visit *www.jigsaw.org/steps.htm*.)

At the end of this activity, students will have an understanding of the content that was presented in *all* of the articles. The jigsaw group is also responsible for generating a list of discussion questions they raised as they shared the content of the articles. Each jigsaw group then makes a short presentation (about 10 minutes) to the whole class summarizing the information from their group discussions and the questions their group generated. As an alternative to a summary of the information about the articles, the teacher might ask each group to present the three most interesting things they learned in their jigsaw experience. The three-pronged approach—expert group, jigsaw group, class presentation—ensures that the students have multiple exposures to the content in a student-centered manner.

Extensions and Cross-Curricular Applications

This activity can launch further studies in organic chemistry, environmental science and engineering, or health studies. After the class has experienced the full jigsaw process and shared group-generated questions, they can branch off into individual research or team research to develop projects to help address these new questions.

For example, a jigsaw group member was concerned about a relative who was pregnant. Her group became very concerned about the health effects of drinking water from plastic bottles because mothers often use bottled water when making formula. The teacher allowed students to search the internet for more information about the quality of drinking water in bottles versus the quality of tap water. A laboratory activity was also developed allowing students to study the concentration of chemicals leaching from plastic bottles.

In addition to helping the class manage science content collaboratively, it should be noted that English language learners are apt to benefit not only from cooperative learning, but also from the purposeful and contextual interaction with new vocabulary embedded within the science content. Science/reading literacy extensions could involve identifying and discussing context clues, inferences, main idea, and author's intent, along with building new vocabulary. Students might also want to research some of the technical terms in the essays.

Essay 1: Plastic and Toys

Products made from plastic enrich our lives and provide many benefits. They and many other valuable products are created from synthetic chemicals, "synthetic" in that they are not found in nature; most plastics are made in the laboratory from petroleum. The synthetic chemicals that compose these products or are used in their production have led to many valuable tools for society—even providing life-saving technologies. However, they can also have a damaging impact on the environment and on human health. Most of these chemicals decompose or break down so slowly they are basically permanent in the environment, or they decompose into other chemicals. Therefore, their effects may be around for a very long time, which means we need to search for alternative technologies, such as biodegradable materials.

The production and use of synthetic chemicals increased greatly around the time of World War II to solve wartime problems. Pesticides such as DDT were developed to control body lice, because body lice could cause typhus. Herbicides were developed to defoliate jungles to provide an advantage during warfare. At the end of the war, the U.S. government helped the petrochemical industry to find markets for their products at home. DDT was used for mosquito and agricultural pest control. Defoliants and insecticides were developed for lawn, garden, and household use. Detergents replaced soaps, and plastic products became abundant (WHO 1979).

Plastic Toys

Toy production boomed with the advent of plastics. In order to make plastics soft and pliable, petroleum-based chemicals called phthalates are added to plastics. Phthalates may be found in shower curtains, upholstery, raincoats, medical devices, and many plastic toys that need to be pliable. The Consumer Product Safety Commission placed regulations on how much phthalate chemicals can exist in toys for sale to children less than three years of age. Toys that babies may put in their mouths that are manufactured in the United States and Canada—teething rings, pacifiers, and soft rattles—are regulated so that no phthalates are added to the plastic. However, toys for older children and toys manufactured in other countries may have phthalates added (Environews 2003). A study in Japan found that high concentrations have been found in some pacifier toys and teething rings. Some governments have proposed a ban on the use of phthalates in all toys for children under the age of three (Sugita et al. 2001).

Living organisms are impacted through the ingestion of these synthetic chemicals as they escape from the plastic items or are released into the environment during their production or use. We are continually exposed to these chemicals in very small quantities—parts per million. We do not know much about the accumulation of the exposure to synthetic chemicals over a lifetime. Some recent studies are telling us that even small quantities can lead to harmful health effects (Welshons et al 2003).

References

Environews. 2003. New arrival: CERHR monograph series on reproductive toxicants. *Environmental Health Perspectives* 111 (13): 696–98.

Sugita, T., K. Hirayma, R. Nino, T. Ishibashi, T. Yamada. 2001. Contents of phthalate in polyvinal chloride toys. *Journal of the Food Hygiene Society of Japan* 42 (1): 48–55.

Welshons, W. V., K. A. Thayer, B. M. Judy, J. A. Taylor, E. M. Curran, and F. S. vom Saal. 2003. Large effects from small exposures. I. Mechanisms for endocrine-disrupting-chemicals with estrogenic activity. *Environmental Health Perspectives* 111 (8): 994–1006.

World Health Organization (WHO). 1979. Environmental health criteria 9: DDT and its derivatives. International Program on Chemical Safety. *www.inchem.org/ documents/ehc/ehc/ehc009.htm.*

Essay 2: Sources of Contamination

Synthetic chemicals, such as those that are used to create plastics, can affect living organisms in several ways. Some of these chemicals disrupt normal hormonal actions. A chemical compound that disrupts normal hormonal actions is called an endocrine-disrupting compound, or EDC. Several compounds, synthetic and natural, can be called EDCs. Synthetic EDCs can mimic estrogen, and in this case they are called xenoestrogens (*xeno* means foreign). These include chemicals sometimes found in detergents, plastics, pesticides, and industrial chemicals. Research has focused on a group of chemicals called phthalates, which are used to make some types of plastics more flexible. Phthalates are found in many common products, such as soft toys, plastic wrap, shower curtains, and medical bags and tubes. Some of the phthalates that can be termed xenoestrogens include the following:

- Di(2-ethylhexyl) phthalate, or DEHP—this synthetic chemical is used in building products, food packaging, children's products, and medical devices

- Bisphenol A—a component of polycarbonate plastic used in electrical equipment, sports safety equipment, protective coatings for cans holding food products, and reusable food and drink containers.

Phthalates Found in Human Urine

Phthalates are fat soluble, and therefore accumulate more easily in fatty tissue, where they may remain for very long periods of time. However, the body also metabolizes phthalates, so exposure can be measured by examining the level of phthalates in urine. Because phthalates are found throughout the environment, it is important to determine if they are concentrating in the human body. Measuring phthalates after the body has processed them yields important information about human exposure to these chemicals and helps to identify potential health risks.

From 1988 to 1994, the Centers for Disease Control ran a study to analyze urine samples from 289 adults ages 20–60. Seven phthalates were identified in the urine samples, and four were found in more than 75% of the samples. The phthalate levels ranged from negligible to levels as high as 15 parts per million (CDC 2000).

Fifteen parts per million is a small percentage; however, the body is very sensitive to hormonal activity, and natural human hormones such as estrogen are often measured in parts per billion. Because these phthalate levels were evident in people's metabolism, and because they can accumulate from exposure to multiple sources, there is cause for further investigation into the source of the chemicals and their effects on the body.

Xenoestrogen Leaching From Plastic Bottles

Several studies show that xenoestrogens can get into drinking water from plastic bottles. Those studies are summarized here:

- A 1993 study found that large plastic bottles used to hold drinking water did contain the xenoestrogen called bisphenol A if they had been exposed to high temperatures or caustic cleaners. The quantity of the chemical was very small, in the parts-per-billion range. However, in laboratory studies with breast cancer cells, even this small quantity was enough to cause the breast cancer cells to increase in number (Krishnan et al. 1993).

- Federal Drug Administration chemists also found that this chemical, bisphenol-A, was leaching from baby bottles that had been heated or were well used. Water in worn and heavily scratched bottles had up to 28 parts per billion of bisphenol-A. Higher levels of the chemical were detected in other baby food products. Although the quantities were small, they are in the same range as those that cause abnormalities in rats (Roloff 1999).

- Two Danish scientists found phthalates in 50% of their samples of baby food and infant formula. From their analysis, the mean intake of a different xenoestrogen, Di(2-ethylhexyl) adipate, or DEHA, for a baby using these products was from 1% to 13% of the tolerable daily intake (Peterson and Breindahl 2000).

Food-Wrap Studies

Chemicals do escape from plastic wrap and other very flexible plastics into foods, and the amount of the chemical contamination increases with time and temperature. Some say these levels of contamination are too low to cause alarm. However, even the plastics industry does not recommend using plastic film when heating foods in the microwave and advises to use only plastic containers marked "microwave safe." Some studies in this area are summarized below.

- A study found levels of a phthalate, especially DEHP, in cream, cheese, and butter at levels of 71 milligrams per kilogram for DEHP and 114 milligrams per kilogram for all phthalates. Then, the researchers exposed cheese to plastic food wrap that contained DEHA. After two hours at 5°C, the level of DEHA was 45 milligrams per kilogram of cheese. After 10 days it increased to 150 milligrams of DEHA per kilogram of cheese (Sharman et al. 1994).

- A similar study tested 49 kinds of plastic film for wrapping foods. Researchers placed olive oil on the plastic wrap for 10 days at 40°C, and then tested the olive oil for traces of DEHA. Substantial amounts of DEHA were found in 42 of the 49 samples, and they were declared illegal for wrapping some foods (Petersen and Breindahl 1998).

- Another study found that different plastic chemicals did exist in a variety of plastic-wrapped foods (cheese, cooked meats, confectionary, meat pies, cake, and sandwiches) from grocery stores (Castle et al. 1998).

References

Castle, L., A. J. Mercer, J. R. Startin, and J. Gilbert. 1998. Migration from plasticized films into foods. *Food Additives and Contaminants* 5: 9–20.

Centers for Disease Control (CDC). 2000. *Levels of seven urinary phthalate metabolites in a human reference population*. Washington, DC: CDC.

Krishnan, A. V., P. Stathis, S. F. Permuth, L. Tokes, and D. Feldman. 1993. Bisphenol-A: Anestrogenic substance is released from polycarbonate flasks during autoclaving. *Endocrinology* 132 (6): 2277–78.

Petersen, J. H., and T. Breindahl. 1998. Specific migration of DEHA from plasticized PVC film: Results from an enforcement campaign. *Food Additives and Contaminants* 15: 600–608.

Peterson, J. H., and T. Breindahl. 2000. Plasticizers in total diet samples, baby food and infant formulae. *Food Additives and Contaminants* 17: 133–41.

Raloff, J. 1999. What's coming out of baby's bottle? *Science News Online* 156 (5). *www.sciencenews. org/sn_arc99/7_24_99/food.htm.*

Sharman, M., W. A. Read, L. Castle, and J. Gilbert. 1994. Levels of di-(2-ethylhexyl) phthalate and total phthalate esters in milk, cream, butter, and cheese. *Food Additives and Contaminants* 11: 375–85.

Essay 3: Effects on Animals' Reproductive Development

Xenoestrogens

Synthetic chemicals such as those found in plastics can affect living organisms in several ways. Some of these chemicals disrupt normal hormonal actions. A chemical compound that disrupts normal hormonal actions is called an endocrine-disrupting compound, or EDC. Several compounds, synthetic and natural, can be called EDCs. Synthetic EDCs can mimic estrogen, and in this case they are called xenoestrogens (*xeno* means foreign). These include chemicals sometimes found in detergents, plastics, pesticides, and industrial chemicals. Research has focused on a group of chemicals called phthalates, which are used to make some types of plastics more pliable. Phthalates are found in many common products, such as soft toys, plastic wrap, shower curtains, and medical bags and tubes. Some of the phthalates that can be termed xenoestrogens and are discussed in this article include the following:

- Di(2-ethylhexyl) phthalate, or DEHP—This synthetic chemical is used in building products, food packaging, children's products, and medical devices

- Bisphenol A—A component of polycarbonate plastic used in electrical equipment, sports safety equipment, protective coatings for cans holding food products, and reusable food and drink containers.

Overview

Cells in animals' reproductive organs have more estrogen receptors and are therefore more sensitive to the effects of these phthalates, especially during fetal and postnatal life. Also, these phthalates accumulate in fat stores in the body and can enter the fetus from accumulated stores in the mother's body. Therefore, there is more concern about the effect of exposure to these xenoestrogens in the fragile states of early development and in their accumulation in mammary tissues.

Reproductive Malformations

Fish from the St. Lawrence River are exposed to xenoestrogens via chemical pollution. This chemical pollution is associated with health problems in the reproductive function in male fish. A Canadian study found that spottail shiners living in the more polluted parts of the river had lower sperm count, and one-third of the fish examined displayed intersex, a condition where ovarian tissue is growing in the testicles (Aravindakshan et al. 2004b). In another Canadian study, rat pups in three groups were fed an additive to their diet.

In the first group, distilled water was added; homogenized fish from a non-contaminated site in the St. Lawrence River were added to the diet of the second group; and the third group was fed homogenized fish from the contaminated site. When the rats reached adulthood, sperm concentrations and sperm motility parameters were significantly decreased in the xenoestrogen group (Aravindakshan et al. 2004a).

- Researchers found that the pesticide pollution in Lake Apopka near Orlando, Florida, could be responsible for sexual deformities in animals, such as smaller penis sizes, and a previous population decline of alligators in the lake. Also, preliminary tests on Florida's biggest lake showed lower testosterone levels and small penis size, as well as altered thyroid hormone levels (Guillette et al. 1994).

- In another study, scientists exposed pregnant mice to several synthetic compounds to see the effect on the development of the second generation's reproductive organs. One of the synthetic chemicals, DEHP, proved to be highly toxic to the reproductive system of male offspring in the study that lasted over several generations of mice. There were high levels of testicular and epididymal abnormalities, including atrophy and agenesis (testes never formed). Rats in the control group showed none of these effects, while 75% of the male rats in the study group had these effects (Gray et al. 2001).

Breast Tissue

Recent laboratory studies give rise to the need for further investigations into the role that xenoestrogens may have in increasing the risk for breast cancer. These studies have found that breast fat and serum lipids of women with breast cancer contain significantly elevated levels of some of the phthalates compared with non-cancer controls (Hoyer et al. 1998).

References

Aravindakshan, J., M. Gregory, D. J. Marcogliese, M. Fournier, and D. G. Cyr. 2004a. Consumption of xenoestrogen-contaminated fish during lactation alters adult male reproductive function. *Toxicological Sciences* 81: 179–89

Aravindakshan, J., V. Paquet, M. Gregory, J. Dufresne, M. Fournier, D. J. Marcogliese, and D. G. Cyr. 2004b. Consequences of xenoestrogen exposure on male reproductive function in spottail shiners (*Notropis hudsonius*). *Toxicological Sciences* 78: 156–65.

Gray, L. E., C. Wolf, C. Lambright, P. Mann, M. Price, R. L. Cooper, and J. Ostby. 2001. Administration of potentially antiadrogenic pesticides and toxic substances during sexual differentiation produces diverse profiles of reproductive malformations in the male rat. *Toxicology and Industrial Health* 15: 94–118.

Guillette, L. J. Jr, T. S. Gross, G. R. Masson, J. M. Matter, H. F. Percival, and A. R. Woodward. 1994. Developmental abnormalities of the gonad and abnormal sex hormone concentrations in juvenile alligators from contaminated and control lakes in Florida. *Environmental Health Perspectives* 102: 680–88.

Hoyer, A. P, P. Grandjean, T. Jorgensen, J. W. Brock, and H. B. Hartvig. 1998. Organochloride exposure and risk of breast cancer. *Lancet* 352 (9143): 1816–20.

Note: The studies listed above are laboratory studies that show a relationship between exposure to the chemical and the health of the organism in the study. Studies that "prove" a connection of exposure to human health are more difficult.

Essay 4: Plastics and Wildlife

This article is shorter than the other three, and should be supplemented with internet research on how plastics impact wildlife and organizations that are trying to reduce that impact and repair some of the damage that has already occurred (see Resources).

Plastics discarded carelessly play a big role in endangering and killing wildlife. This is particularly evident in marine environments. An estimated two million seabirds and 100,000 marine mammals are killed every year by lost or discarded plastics (Conner and O'Dell 1988). Fishing boats and commercial boats, misuse of rivers and streams, sewage waste, illegal dumping, and careless people contribute to the pollution. The litter can also result from storms or shipwrecks. Marine litter is found in the water column, along the shore, on the seabed, and floating on the surface of the water. Almost 90% of this floating litter is plastic.

Plastics and other marine litter affect animals in several ways. Often the animals accidentally ingest the plastics, as the plastics appear to the animals like food or are attached to something they are eating. Turtles and fish have been found with tangled masses of fishing line in their stomachs, and plastic items are commonly found in the stomachs of many water and shore animals.

In addition, plastics and other marine debris kill and endanger animals through entanglement. Some plastic articles last hundreds of years with-out degrading, and they are accumulating in the ocean as more and more are carelessly handled or discarded. The main cause of wildlife entanglement is abandoned fishing line. The effects of entanglement include infection, amputation of limbs, digestive blockage, and poisoning from the chemicals in the plastics (Hutchinson and Simmonds 1992).

References

Conner, D. K., and R. O'Dell. 1988. The tightening net of marine plastics pollution. *Environment* 30 (1): 16–17.

Hutchinson, J., and M. Simmonds. 1992. Escalation of threats to marine turtles. *Oryx* 26 (22): 95–102.

Internet Resources

Pollution of the Ocean by Plastic and Trash
www.water encyclopedia.com/Po-Re/Pollution-of-the-Ocean-by-Plastic-and-Trash.html

What You Can Do About Plastic Pollution
www.algalita.org/pdf/What-you-can-do.pdf

Plastic Pollution in Pacific Waters
www.the environmentalblog.org/2007/10/plastic-pollution-in-pacific-waters.html

Ocean Plastic Pollution and How You Can Help
http://oceans.greenpeace.org/en/the-expedition/news/trashing-our-oceans

World's Oceans Face Plastic Pollution Problem
www.pbs.org/newshour/extra/video/blog/2008/11/worlds_oceans_face_plastic_pol.html

CHAPTER 12

Designed By Nature

Exploring Linear and Circular Life Cycles

By Sandra Rutherford, Bonnie Wylo, Peggy Liggit, and Susan Santone ///////////////////////////

Designed by Nature is a series of six sequential lessons that help middle school students explore the environmental impacts of producing and disposing of everyday items, and the role of citizens, consumers, and businesses in promoting sustainable product design. The activities, which were developed by faculty at Eastern Michigan University, incorporate the 5E learning model. The steps of this model—engage, explore, explain, elaborate, and evaluate—allow students to explore the environmental implications of design and manufacturing decisions; examine real-world success stories of products that minimize pollution, use less energy, and facilitate easy reuse and recycling; develop ways to design, assess, and select environmentally friendly products; and learn about careers in engineering, design, business, and related fields. Specifically, students will learn the following:

- The production, use, and disposal of an item are referred to as its *life cycle*. The steps of a product life cycle include the extraction of resources, transformation or processing of the raw material into the final product, consumption, and disposal or recycle. Each stage requires energy and resource inputs, and creates outputs.

- Within the product life cycle, manufacturers are responsible for extraction and production and consumers are responsible for usage and disposal.

- The production, use, and disposal of everyday products create social and environmental consequences.

- Changing manufacturing practices involves actions by consumers, citizens, and policy makers.

The lessons guide students through a typical product's life cycle—from producer to consumer to landfill—and then explain that a Designed by Nature product is one that breaks this cycle by being reusable, recyclable, or biodegradable. The six lessons in the series are as follows:

- Lesson 1: Introduction to Product Life Cycles

- Lesson 2: How "Green" Are Your Clothes?

- Lesson 3: A Decomposer's Dilemma (see Activity sheet)

- Lesson 4: Our Trash Never Goes Away!

- Lesson 5: Price Versus Cost

- Lesson 6: Design Challenge

After working through these lessons, students will hopefully become more environmentally and socially conscious of the choices they have when they go to buy a product. Even a simple product like a T-shirt or a bag of potato chips can impact the environment and the people who make the product. In this way and many others, the Designed by Nature program demonstrates how science affects us personally and our society in a cohesive, practical, and fun curriculum. For more information about the lessons, professional development, and to download the free activities, visit *www.emich. edu/biology/DbyN/lessons.html*.

Reference

Applehof, M. 1997. *Worms eat my garbage*. Kalamazoo, MI: Flower Press.

A Sample of Designed by Nature's Lesson 3: A Decomposer's Dilemma

*The **Engage** part of this lesson starts with different kinds of packing peanuts.*

Hold up a Styrofoam packing peanut and ask students to identify it. What is the purpose of this product? What is it made of? Pass a few of them around. Are they biodegradable? Can they decompose? What do we mean by biodegradable and decomposable? Take answers from students to pre-assess their knowledge of these terms and create a casual "working definition" of these terms, depending on what they say. Then ask how this Styrofoam peanut can be disposed of. Some students might just say you throw them away, but some might know you can take them to a recycling station (some recycling centers do not accept Styrofoam, but mailing stores will accept donations of clean Styrofoam and other packing materials).

Hold up a cornstarch packing peanut. Ask students to identify it. What is it made of? (If they don't know, try not to tell them…they may also

think it is Styrofoam; or tell them it is cornstarch, and go from there). Pass a few around. Are they biodegradable? Are they decomposable? Then the teacher can either EAT this cornstarch packing peanut, or put it in a glass of warm water and stir it until it begins to dissolve (preferred). Try the same thing with the Styrofoam peanut; it won't dissolve. (Don't try to eat it.)

*The **Explore** part of this lesson uses what students have seen in the Engage part to inquire about decomposition rates.*

Next, have students set up a worm bin outside the classroom to test whether packing peanuts are biodegradable (see Applehof 1997 for worm bin building instructions; see "Scope on Safety: Debugging Safely" for safety and proper use in the classroom at *www.nsta.org/middleschool*). Place the worm bin on the ground and have students circle around. Pick several students and have them put on nonlatex gloves. Instruct them to move around the contents of the worm bin so all students can see the worms, shredded newspaper, and kitchen scraps in various stages of decay. Have students add a few cornstarch

---

FIGURE 12.1.

Decomposition worm-bin observation sheet

Date and time	Styrofoam packing peanut observations	Cornstarch packing peanut observations

---

and Styrofoam packing peanuts to the premade worm compost bin. Let students observe the worm bin and chart what happens after one minute, one and a half hours, two weeks, and three weeks. They can record their data in the chart in Figure 12.1 (p. 85).

*The **Explain** part of this lesson uses the life cycles shown in Figure 12.2. The most important element to cover in this Explain section is the relationship between the material a product is made from and the ability of a decomposing organism to break down those materials.*

Bacteria, fungi (mold and mushrooms), and invertebrates such as worms digest dead organisms; they have special enzymes that break apart tissues and cells into their simple chemical elements. These elements get returned back into the air and soil to be reused again by nature. Thus, these products have a "Designed by Nature" or circular life cycle. The rate of decomposition is dependent upon environmental factors such as temperature, availability of oxygen, and humidity or dryness. The dead organism's size and weight also affect how long it takes to decompose.

To apply the facts about decomposition to human-made products, products that are composed of plant and animal materials are decayed by decomposing organisms; products that contain human-made products (synthetic fibers, glass, and plastic) cannot be attacked by decomposers. The materials that make up these products cannot be broken down by the digestive enzymes of these organisms. These products have a linear life—the used-up product dead-ends at a landfill.

The following websites contain useful information for you as a teacher:

- *www.bottlebiology.org/investigations/ decomp_main.html*
- *www.bottlebiology.org/investigations/decomp_ bkgreading.html*
- *www.herbarium.usu.edu/fungi/FunFacts/Decay.htm*

*In the **Elaborate** part of this lesson students will apply what they have learned about products that are biodegradable or not. A portion of the handout is on page 87.*

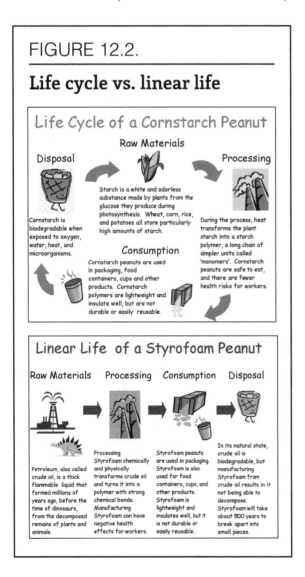

FIGURE 12.2.

Life cycle vs. linear life

Life Cycle of a Cornstarch Peanut

Raw Materials

Disposal   Processing

Starch is a white and odorless substance made by plants from the glucose they produce during photosynthesis. Wheat, corn, rice, and potatoes all store particularly high amounts of starch.

Cornstarch is biodegradable when exposed to oxygen, water, heat, and microorganisms.

During the process, heat transforms the plant starch into a starch polymer, a long chain of simpler units called 'monomers'. Cornstarch peanuts are safe to eat, and there are fewer health risks for workers.

Consumption

Cornstarch peanuts are used in packaging, food containers, cups and other products. Cornstarch polymers are lightweight and insulate well, but are not durable or easily reusable.

Linear Life of a Styrofoam Peanut

Raw Materials   Processing   Consumption   Disposal

Petroleum, also called crude oil, is a thick flammable liquid that formed millions of years ago, before the time of dinosaurs, from the decomposed remains of plants and animals.

Processing Styrofoam chemically and physically transforms crude oil and turns it into a polymer with strong chemical bonds. Manufacturing Styrofoam can have negative health effects for workers.

Styrofoam peanuts are used in packaging. Styrofoam is also used for food containers, cups, and other products. Styrofoam is lightweight and insulates well, but it is not durable or easily reusable.

In its natural state, crude oil is biodegradable, but manufacturing Styrofoam from crude oil results in it not being able to decompose. Styrofoam will take about 500 years to break apart into small pieces.

Categorization of Clothing Elaboration Activity

Consider everything on your person and fill in the information below for TWO items only. For instance, if you are wearing a polyester sports jersey, the polyester and nylon components are made from processed petroleum that is not biodegradable. For a pair of wool socks, the materials are probably wool and cotton and nylon, where the wool and cotton are biodegradable, but the nylon is not. Probable disposal for old socks would be "landfill" (however, if they are not too far gone, you might give them away for "reuse").

Item #1: _____
Raw materials:

Which of these raw materials are biodegradable?

How will you likely dispose of this item?

*The **Evaluate** portion of the lesson is shown below.*

Postassessment for Lesson 3: A Decomposer's Dilemma

"Show Us What You Learned" questions:

List three items you can see in this room that you think are biodegradable. (3 points)

1.

2.

3.

List three items you can see in this room that you think are nonbiodegradable. (3 points)

1.

2.

3.

Do you have a compost pile, bin, or jar at home?

Yes   No

Do you know anyone with a compost pile?

Yes   No

Look outside. What can you see out there that is decomposing right now?

List a product (or part of a product) that would be considered as having a Designed by Nature circular life cycle.

List a product that would be considered as having a linear life cycle.

CHAPTER 13

The Sport-Utility Vehicle
Debating Fuel-Economy Standards in Thermodynamics

By Shannon Mayer //

The environmental challenges we face in the 21st century are scientifically complex, multidisciplinary problems. The next generation of scientists and engineers will be the repository of the technical expertise, as well as the source of technological innovation; it is essential that they be key contributors in the debate and formulation of national and international science policy. An exciting challenge we have as science educators is to investigate how to develop in our science and engineering students the ability to ask the types of questions, both personally and on a larger scale, that maximize constructive involvement in these critical conversations.

This paper describes a debate about national fuel-economy standards for sport-utility vehicles (SUVs) used as a foundation for exploring a public policy issue in the physical science classroom. The subject of automobile fuel economy benefits from a familiarity with thermodynamics, specifically heat engines, and is therefore applicable to a broad range of courses found in the disciplines of physics, chemistry, and engineering. This topic could also be discussed, with somewhat less techni-cal sophistication, in a physical science course for nonscience majors.

Thermal and Statistical Physics

This discussion of SUV fuel-economy standards has been used in Thermal and Statistical Physics, a semester-long, upper-division course for physics majors and minors. Typical enrollment is approximately 10 students. The primary course objective is to examine the fundamental principles of thermal physics from both statistical and classical viewpoints and to apply these principles to thermodynamic systems. A secondary goal of the course is to provide students with activities in which they can apply their developing technical expertise in thermodynamics to societally relevant issues. Practical curricula related to these topics are incorporated into the course, supplementing the traditional course materials (Mayer 2007; Mayer 2006).

The SUV fuel-economy debate was used following a discussion of heat engines. The traditional course content included a description of the heat engine, the Carnot cycle, and a description and comparison of different theoretical engine designs (internal combustion [i.e., Otto cycle] and diesel) and their efficiencies. Students were also given a brief introduction to the science of hydrogen fuel cells. Material related to the heat engine was supplemented with curricula

related to automobile efficiencies and a description of some advanced automotive technologies (described below); this material provided both a practical and interesting application of the topic of heat engines and a technical foundation for the fuel-economy debate.

Background: U.S. Fuel-Economy Standards

In response to the Arab oil embargo of 1973–1974, Congress passed the Energy Policy and Conservation Act in 1975, which, among other things, established corporate average fuel economy (CAFE) standards for new passenger automobiles sold in the United States. New car fuel economy

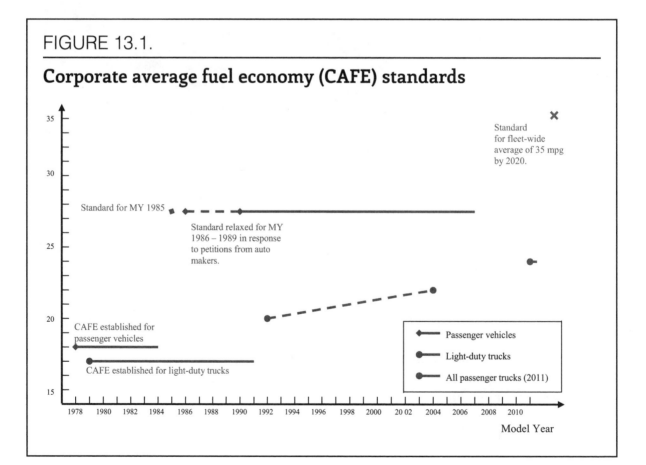

FIGURE 13.1.

Corporate average fuel economy (CAFE) standards

had decreased from a fleet average of 14.8 mpg in 1967 to 12.9 mpg in 1974 (Bamberger 2002). The 1975 CAFE legislation called for a fuel-economy standard of 18 mpg for new passenger vehicles in model year (MY) 1978 and a rise to 27.5 mpg by MY 1985. The CAFE standard remained at this level for MY 2007.

Fuel-economy standards were later set separately for light-duty trucks (gross vehicle weight rating [GVWR] less than 8,500 lb.). A summary of fuel-economy standards for both new passenger cars and light-duty trucks is shown in Figure 13.1 (NHTSA 2007). The automobile fuel economy for a sample of MY 2007 vehicles is shown for reference in Table 13.1. Light-duty trucks that exceed 8,500 GVWR are not required to comply with CAFE standards. Manufacturer compliance with the standards is determined by calculating a sales-weighted average fuel economy for the company's product line. Manufacturers that fail to comply with the standards incur a penalty of $5 for every 0.1 mpg below the standard, multiplied by the number of new vehicles in the manufacturer's fleet for that model year. Auto manufacturers paid approximately $475 million in civil penalties between 1983 and 1998 for noncompliance (Bamberger 2002).

Sport-utility vehicles are classified in the same category as light trucks, allowing them to operate at lower fuel economy than standard passenger vehicles. In 1988, the light-truck classification made up roughly 30% of the vehicle fleet, compared with 45% in 2000. The significant increase in popularity of SUVs in the 1990s contributed to a reduction in overall average fuel economy since the mid-1980s (Bamberger 2002). The separate classification for light-duty trucks, and the corresponding lower-mileage standard for SUVs, is often referred to as the "light-truck loophole." Recent legislation has sought to rectify this dispar-

ity by increasing the fuel-economy standard for light trucks and reforming the structure of CAFE (USDOT 2006). Under the reformed structure, set to go into effect with MY 2011 vehicles, standards will be set for all passenger trucks according to vehicle size. Incremental target levels will scale based on vehicle footprint, with an overall requirement of 24 mpg for all passenger trucks. Moreover, beginning in MY 2011, large SUVs and trucks with GVWR between 8,500 and 10,000 lb. will be included in the CAFE requirements for the first time. A bill passed by Congress in December 2007 further increased the CAFE standards, beyond the 2011 reform, to a fleetwide average of 35 mpg by 2020 (Simon 2007).

Background: Automobile Fuel Economy

The internal combustion engine is based on the idealized four-stroke cycle (Otto cycle) shown in Figure 13.2 on page 93 (for example, Schroeder 2000; Tipler and Mosca 2004). The working substance is a mixture of air and vaporized gasoline, which is injected into a cylinder and adiabatically compressed by a piston. During the adiabatic compression stroke, PV^γ is constant, where P is the pressure, V is the volume, and γ is the ratio of the heat capacity at constant pressure to the heat capacity at constant volume (e.g., $\gamma = 1.4$ for a diatomic gas). During the ignition stroke, the air-gas mixture is ignited by the spark plug, producing an increase in temperature, and therefore the pressure, as the volume is held constant. During the power stroke, the high-temperature gas pushes the piston outward, producing mechanical work. During the exhaust cycle, the hot exhaust gas is expelled at constant volume (in a real engine, the exhaust gas leaves the engine and is replaced by a new air-fuel mixture during this step), returning the engine to the beginning of the cycle.

TABLE 13.1.

Automobile fuel economy: Model year 2007

Model	Specifications	EPA MPG (city)	EPA MPG (hwy)
Sedans	**Automatic**		
Honda Accord	4 cyl, 2.3 L	24	34
Toyota Camry	4 cyl, 2.3 L	24	33
Volkswagen Passat	4 cyl, 2 L	22	31
Pontiac Grand Prix	6 cyl, 3.8 L	20	30
Ford Taurus	6 cyl, 3 L	20	27
Pickups			
Ford Ranger	2WD, 4 cyl, 2.3 L	24 (manual)	29 (manual)
		21 (automatic)	26 (automatic)
	2WD, 6 cyl, 3 L	18 (manual)	23 (manual)
Mazda B2300	2WD, 4 cyl, 2.3 L	24 (manual)	29 (manual)
		21 (automatic)	26 (automatic)
Toyota Tacoma	2WD, 4 cyl, 2.7 L	23 (manual)	28 (manual)
		21 (automatic)	27 (automatic)
	4WD, 6 cyl, 2.7 L	19 (manual)	23 (manual)
	4WD, 6 cyl, 4 L	19 (automatic)	22 (automatic)
Dodge Dakota	2WD, 6 cyl, 3.7 L	16 (automatic)	22 (automatic)
SUVs	**Automatic**		
Jeep Patriot	2WD, 4 cyl, 2 L	26	30
Honda CR-V	2WD, 4 cyl, 2.4 L	23	30
Toyota Highlander	2WD, 4 cyl, 2.4 L	22	28
Ford Escape	4WD, 6 cyl, 3 L	20	24
Honda Pilot	4WD, 6 cyl, 3.5 L	18	24
Nissan Pathfinder	2WD, 6 cyl, 4 L	16	23
GMC Yukon 1500	2WD, 8 cyl, 5.3 L	16	22
Chevrolet Suburban 1500	2WD, 8 cyl, 5.3 L	15	21
Ford Explorer	2WD, 6 cyl, 4 L	15	21
Lincoln Navigator	2WD, 8 cyl, 5.4 L	13	18
Mercedes-Benz 655 AMG	8 cyl, 5.4 L	12	14

(SOURCE: USDOE 2007A.)

The efficiency, ε, of an engine is defined to be the ratio of the work done by the engine to the heat absorbed from a high-temperature reservoir. For an automobile engine there is not a high-temperature reservoir as such, but rather thermal energy is generated internally by the burning of the fuel. The automobile engine efficiency is then the work (W) produced by the power stroke divided by the heat absorbed (Q_H) during the ignition cycle. For the Otto cycle shown, the ideal efficiency is

$$\varepsilon = \frac{W}{Q_H} = 1 - \frac{1}{r^{\gamma-1}}$$

where the compression ratio $r = V_1/V_2$ is the ratio of the volume at the beginning and ending of the compression stroke. The compression ratio for an automobile engine is typically 8 to 10, corresponding to a theoretical efficiency of approximately 0.56 to 0.6 (assuming γ ≈1.4). Factors such as friction, heat loss, and incomplete combustion of gases reduce the achievable efficiency. Automobile efficiencies typically range from 20% to 30%, with the best internal combustion engines operating near 50% efficiency (Tipler and Mosca 2004).

The optimal efficiency of an automobile depends upon a number of factors in addition to the engine efficiency. Primary factors include drivetrain efficiency, aerodynamic drag, idle time, vehicle weight (i.e., inertia), and accessory energy consumption (air conditioning, windshield wipers, and so on). The approximate contribution of these factors to overall automobile efficiency is summarized in Table 13.2 (p. 95). The table also lists a sample of advanced automotive technologies aimed at improving efficiency in these areas along with their projected improvement (USDOE 2007b). Several of these advanced technologies are described briefly below. Depending on the goals of a particular course and the interests of students, any number of automotive innovations could be researched and incorporated into the curriculum.

Subjective factors such as personal driving style and vehicle maintenance also impact automobile efficiency, though the EPA determines

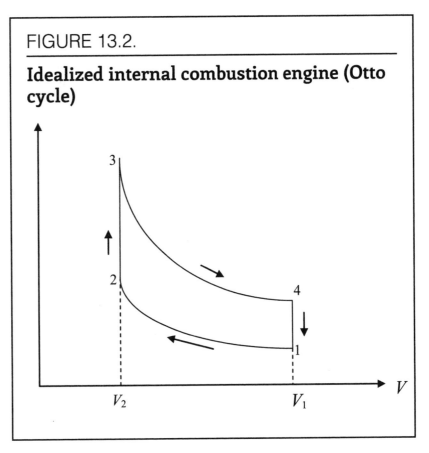

FIGURE 13.2.

Idealized internal combustion engine (Otto cycle)

vehicle fuel economy based on ideal conditions. Several of these subjective factors, along with techniques for improving efficiency in these areas, are included for interest in Table 13.2.

The primary loss mechanism in an ideal automobile is engine inefficiency. Equation 1 shows that engine efficiency can be improved by increasing the compression ratio. One way to accomplish this is by using a technique called *direct fuel injection*. As the name suggests, fuel is injected directly into the combustion cylinder rather than being mixed with air before being pumped into the cylinder. Direct fuel injection allows for more precise control over the amount of fuel injected, the shape of the mist, and the injection timing; as a result, fuel intake is more efficient and a higher compression ratio can be achieved. Direct fuel injection has been implemented in a number of vehicles, including various models made by Mitsubishi, Toyota, BMW, GM, and Isuzu.

Valves are used to control the intake and exhaust from the pistons in an engine. The timing of the valve opening and closing and the amount of lift are important factors in determining engine efficiency. The optimum timing depends on the engine speed. In a conventional system, the valve timing and amount of lift are fixed, independent of the engine speed, at some intermediate value. With a *variable valve timing and lift* (VVT&L) system, the timing and lift are automatically optimized to the speed of the engine. Variable valve time is available on the 2008 Honda Accord, the 2007 Mazda 6, and other select vehicles.

The engine idle time associated with waiting at lights or being stopped in traffic is also a significant factor in overall automobile efficiency. New technologies such as the *integrated starter/generator* (ISG) seek to reduce this loss by automatically turning off the engine when the vehicle is stopped and instantaneously turning it back on

upon actuation of the accelerator (USDOE 2007b). This technology is currently under development.

Drivetrain inefficiency results in engine power not being efficiently transferred into translational motion of the automobile. *Automated manual transmission* (AMT) combines the fuel efficiency of a manual transmission with the convenience of an automatic transmission. With AMT, the shifting of gears is accomplished electronically, rather than manually. One vehicle available with the AMT feature is the Audi R8.

Automobile efficiency is a multifaceted, technical problem. The class discussion regarding engine operation and automobile efficiency provides students with a technical foundation from which they can be informed participants in the fuel-economy debate.

The In-Class Discussion

To prepare for the science policy discussion, students' technical foundation was supplemented with an excerpt of testimony presented at a U.S. Senate subcommittee hearing in December 2001 (Cooper 2002) and a few short articles from the popular press (*Oregonian* 2005; *Chronicle of Higher Education* 2005; Lauer 2005; Scully 2004). The pro/con summaries used in this activity are available in a report by *CQ Researcher* on the broader topic of energy security. (The *CQ Researcher* provides original, in-depth reporting and analysis on current events and covers topics such as health, education, the economy, the environment, and technology. Their single-topic reports provide researchers with a thorough introduction, background, and up-to-date overview of a given topic. Included in each report are succinct pro/con statements from representatives holding opposing viewpoints. In the report on energy security, the pro/con positions regarding raising fuel-economy standards are summaries

TABLE 13.2.

Automobile efficiency data from the U.S. Department of Energy

Energy-loss mechanism	Energy consumption	Advanced automotive technology	Average efficiency increase (projected or actual)
Engine losses (internal combustion engine)	62.4%	Variable valve timing and lift (VVT&L) Turbo-charging Direct fuel injection Cylinder deactivation	5% 7.5% 12% 7.5%
Standby/idle	17.2%	Integrated starter/generator (ISG)	8%
Drivetrain inefficiency	5.6%	Automated manual transmission (AMT) Continuously variable transmission (CVT)	7% 6%
Accessory energy consumption	2.2%		
Total loss	87.4%		
How is the remaining energy used?	**Energy consumption**	**Improving efficiency**	**Fuel-economy benefit**
Aerodynamic drag	2.6%		
Rolling friction	4.2%		
Overcoming inertia/ braking losses	5.8%	Minimize carrying excess weight in vehicle	1%–2% per 100 lb.
Total loss	12.6%		
Variability in automobile efficiency	**Excess energy consumption**	**Improving efficiency**	**Potential fuel-economy benefit**
Driving practices	Variable	Avoid aggressive driving (speeding up, rapid acceleration, and so on). Observe the speed limit (gas mileage decreases significantly for speeds greater than 60 mph). Avoid excess idle time. Use cruise control.	5%–33% 7%–23%
Automobile maintenance	Variable	Tune up engine. Maintain air- and fuel-filter maintenance. Maintain proper tire inflation. Use recommended grade of motor oil.	4% up to 10% up to 3% 1%–2%

(SOURCE: USDOE 2007B.)

> ## FIGURE 13.3.
>
> ## Questions used in preparation for the in-class discussion
>
> 1. Summarize the primary argument in favor of increasing fuel-economy standards.
>
> 2. Summarize the primary argument in opposition to increasing fuel-economy standards.
>
> 3. Which argument do you find most compelling and why?
>
> 4. What do you see as the most significant obstacle faced by the position promoted by the Union of Concerned Scientists? By General Motors?
>
> 5. Does either side have a bias that might influence the opinion that it presents?
>
> 6. What biases do you bring to the table that might influence your decision to support one side or the other? Are these biases *good* or *bad*?
>
> 7. What questions would you need to research in order to develop a better scientifically informed position? How would you go about answering those questions? Select one question to research further.
>
> 8. What lengths should we go to in order to reduce our dependence on foreign oil? What would you be willing to give up?
>
> 9. What factors are important to consider as we balance issues of business with issues of public good?

of testimony from the U.S. Senate subcommittee hearing. For complete details of the hearing see U.S. Senate 2001). The Senate testimony, which provided the primary framework for the in-class discussion, presented succinct pro and con arguments regarding the question of whether automobile (not exclusively SUV) fuel-economy standards should be tightened in order to reduce dependence on foreign oil. The pro argument was put forth by the Union of Concerned Scientists. The con argument was presented by General Motors Corporation. Prior to the in-class discussion, students were provided with several questions to consider related to the Senate testimony and SUV fuel-economy standards (see Figure 13.3). The questions led students to summarize the primary argument in favor of and opposed to raising fuel-economy standards, to identify the biases (good or bad) that both sides—and they themselves—might bring to the debate, and to identify and research issues brought up in the Senate testimonies and other readings that would benefit from further research in their development of a more informed opinion.

Following students' preparation outside of class, we convened to discuss the issue. The discussion took place during one class period, with the instructor serving as the moderator and each student contributing to the discussion. Students summarized the compelling points of the arguments presented in the pro/con summaries, provided further arguments specific to the standards for SUVs, discussed evaluating technical information, questioned how bias (social, political, financial) might influence the presented opinions, and explored how their recently acquired scientific expertise would help inform their opinion regarding the issue. Supplemental research that students conducted prior to the discussion also informed the dialogue; examples include investigation of the status of hydrogen fuel cells, the impact of fuel-economy standards on road safety, practical (engineering) feasibility of implementing higher fuel-economy standards, and possible economic impact of higher fuel-economy standards.

Outcomes

The primary learning objective for this activity was the development of students' ability to use scientific expertise and technical data to critically analyze arguments about a social/environmental issue. More generally, I wanted to introduce students to the arena of science policy and allow them to discover how technical expertise can enhance their ability to develop an informed opinion regarding a societally important issue.

Assessment of student learning was based on (1) instructor assessment of and student feedback on the classroom discussion, (2) an individual writing assignment following the in-class discussion, and (3) assessment of a follow-up activity in which students constructed their own pro/con arguments about another science policy issue.

Students were active and enthusiastic participants in the fuel-economy discussion. This is a notable outcome, as research has shown that student engagement enhances student learning (Hake 1998). Many students reported that this was their first opportunity to engage in this type of conversation/debate in a physical science course and they valued the opportunity to apply their science training to a practical question. Several students also recognized and commented on their own difficulty in reconciling their scientific opinion with their personal preferences.

One significant discovery made by students was that their technical expertise did enhance their ability to understand the issues and to develop an informed position. For example, one argument

against raising the fuel-economy standard was based on testimony by General Motors that it was "investing significant engineering resources to create a completely revolutionary technical capability," namely vehicles powered by hydrogen fuel cells, and that new legislation would be an obstacle to achieving this goal (Cooper 2002). GM states in its comments that "[h]ydrogen fuel made from renewable sources of energy can be used to power fuel-cell vehicles that are more than twice as energy efficient as today's vehicles and emit only pure water" (Cooper 2002). Students' background regarding some of the technical aspects of fuel-cell technology gave them a foundation from which to carefully evaluate this position. Moreover, their understanding of the inherent challenges (such as the lack of a readily available source of hydrogen), by-products, and processes involved in fuel-cell technology provided them with technical expertise that they could draw upon to support their arguments (pro or con).

Following the in-class discussion, students prepared written responses to the questions from the pre-discussion guide, and included in their response a summary of any further research they had conducted. The writing assignment challenged students to intentionally reflect on the activity and to carefully articulate their arguments. The writing assignment was graded individually with assessment based on the clarity of students' responses, the quality of their writing, and the constructive use of additional research to support their position.

The fuel-economy discussion also provided a foundation for a public policy debate carried out by students later in the semester. In the follow-up debate, students were assigned to teams and given the challenge to construct and present pro/con arguments regarding the proposal for an offshore wind farm on public lands in Nantucket

Sound. The opportunities presented by the fuel-economy activity—to critically analyze arguments, to conduct technical research to support a position on a public policy issue, and to clearly articulate a position—provided students with a valuable framework from which to prepare for the wind-power debate. Details about the wind-power debate can be found in Mayer 2007. Student responses to the wind-power debate, SUV activity, and other environmentally related course activities are resummarized from that article below.

Written comments, coupled with my conversations with students, indicate that the science policy–related course activities did provide students with the opportunity to think *intentionally* about physics and its relation to society. Students reported that they appreciated the opportunity to bring their scientific knowledge to bear on relevant societal and environmental issues. Moreover, they indicated that this component of the course aided them in making connections between their major and their core liberal arts courses. One student commented, "This particular course brought together many ideas and concepts that had been presented, but not linked, in my previous five years of physics classes. It was a good 'last course.'" Another said, "I thought the course was great, especially the effort to relate material to our society/politics/environment." Another student wrote, "Liberal arts additions to the course (environmental essays) were very effective."

Conclusion

Whether physical science and engineering graduates spend their careers designing technological innovations, conducting basic research, developing national science policy, or simply engaging in the world as citizens, the science community has an opportunity and an obligation to participate in the public arena. Moreover, scientists and engi-

neers have the unique ability to bring their scientific training and knowledge to bear in addressing critical environmental issues. This discussion about national fuel-economy standards provides students with an opportunity to draw connections between their developing scientific expertise and an interesting and current environmental public policy issue. The process of asking questions and exploring intersections provided by this activity may encourage students to develop the habits of thought that will motivate and enable them to be involved participants in the important conversations of their generation.

References

Bamberger, R. 2002. Automobile and light truck fuel economy: CAFE standards. *Almanac of Policy Issues* (September). *www.policyal manac.org/ environment/archive/crs_cafe_standards.shtml*.

Cooper, M. H. 2002. Energy security. *CQ Researcher Online* 12: 73–96. *http://library.cqpress.com/cqre searcher/cqresrre2002020100*.

Chronicle of Higher Education. 2005. Fear-driven SUVs. November 4: B4.

Hake, R. R. 1998. Interactive engagement versus traditional methods: A six thousand student survey of mechanics test data for introductory physics courses. *American Journal of Physics* 66 (1): 64–74.

Lauer, J. 2005. Driven to extremes: Fear of crime and the rise of the sport utility vehicle in the United States. *Crime, Media, Culture* 1 (2): 149–68.

Mayer, S. 2006. Intersections: Exploring social justice issues in the physical sciences. In *Teaching, faith, and service: The foundation of freedom*, eds. W. Hund and M. M. Hogan, 215–24. Portland, OR: University of Portland Press.

Mayer, S. 2007. Cape wind: A public policy debate for the physical sciences. *Journal of College Science Teaching* 35 (7): 24–27.

National Highway Traffic Safety Administration (NHTSA). 2007. Vehicles and equipment: CAFE standards. *www.nhtsa.dot.gov/cars/rules/cafe/index.htm.*

Oregonian. 2005. Lead, follow or get out of the way. August 26.

Schroeder, D. V. 2000. *Thermal physics.* San Francisco: Addison Wesley Longman.

Scully, M. G. 2004. The end of easy oil. *Chronicle of Higher Education* October 1: B11.

Simon, R. 2007. Energy bill boosts fuel economy standards. *Los Angeles Times.* December 19. Available online at *www.latimes.com/news/nationworld/la-na-energy19dec19,0,1969731.story?coll=la-home-center.*

Tipler, P. A., and G. Mosca. 2004. *Physics for scientists and engineers.* 5th ed. New York: W.H. Freeman.

U.S. Congress. Senate. 2001. Committee on Commerce, Science, and Transportation. Corporate Average Fuel Economy (CAFE) Reform Hearing before the Committee on Commerce, Science, and Transportation United States Senate, 107th Congress, 1st sess., Dec. 6, 2001. Washington, DC: U.S. Government Printing Office, 2005. (Y4.C 73/7:S.HRG.107-1137).

U.S. Department of Energy (USDOE). 2007a. Find and compare cars: Search by market class. *www.fueleconomy.gov/feg/byclass.htm.*

U.S. Department of Energy (USDOE). 2007b. Hybrids, diesels, alt fuels, etc.: Advanced technologies and energy efficiency. *www.fuele conomy.gov/feg/atv.shtml.*

U.S. Department of Transportation (USDOT). 2006. New light truck economy standards to save 10.7 billion gallons of fuel, include largest SUVs for first time ever. *www.dot.gov/affairs/cafe032906.htm.*

PART TWO

Content Area Activities

CHAPTER 14

Design Challenges Are "ELL-ementary"

English Language Learners Express Understanding Through Language and Actions

By Nancy Yocom de Romero, Pat Slater, and Carolyn DeCristofano ////////////////////////////////

Classmates' excited chatter about a science concept overwhelms a student with trouble processing spoken language. In another classroom, a teacher labors to teach about material properties to English language learners. "Odio la ciencia! (I hate science)," mutters one child, who loves the hands-on work but lacks sufficient English to feel successful.

As teachers in a school with large bilingual and special education populations, we constantly seek new ways to help students access science concepts. And we are not alone. Elementary teachers across the nation are facing similar questions: How do we help our special needs students and English learners understand challenging, standards-based science content while they are developing English language skills?

Through our work as pilot teachers for the Engineering is Elementary (EiE) program developed by the Museum of Science in Boston (see Internet Resources), we've discovered an exciting answer: carefully conceived design challenges. Design challenges—using science knowledge to design, create, and test some thing or process—encourage the development and use of science concepts and English language in contexts that students find meaningful. In design challenges, students can work with content without relying heavily on language and express their science theories in actions, not just words.

We tested EiE's *Materials Engineering: Designing Walls* unit separately—each in our respective classrooms: a fourth-grade bilingual group studying rocks, minerals, and earth materials, and a second-grade inclusion class studying plants and soil—and had terrific success. We each completed our units in about one and one-half weeks (seven 45-minute class periods). The following is a description of our experiences.

Designing Walls

The unit's design challenge—designing a wall to be built from earth materials that could meet specific criteria—engaged students meaningfully. First introduced in the context of a storybook and later pursued as a real class problem, the unit provided the motivation, thematic focus, and momentum that propelled students through challenging content—new concepts, words, and communication skills. Figure 14.1 is an overview of the four-lesson unit and how it relates to the EiE curriculum framework.

Before we got started, we gave an overview of engineers and technology with the objective of helping students understand that engineers design technologies—and that technologies include any object intended to solve a problem. In small groups, students received mystery objects (identified as technology) hidden in paper bags. Upon opening the bags, students discovered that technology includes items as familiar as toothbrushes, paper cups, and other everyday objects. Students then discussed among themselves where the objects came from and what sorts of problems these objects are intended to solve.

Although this was a simple activity, we were impressed that students picked up on the words and concepts of engineering and technology very quickly. Immediately, they seemed to feel empowered by these additions to their vocabulary. Outside of the lesson context, for example, the second graders would refer to dif-

---

FIGURE 14.1

Engineering unit outline

Unit Lesson	EiE design cchallenge framework	Materials engineering: designing walls
1	Storybook introduces a problem that students investigate and for which they design a solution.	Students read Yi Min's Great Wall.
2	Activity-based exploration of technology and engineering	Students choose among samples of different materials to perform certain tasks, identifying useful properties.
3	Science exploration of concepts related to design challenge	Students explore and describe properties of different earth materials.
4	Students solve design challenge related to storybook using creativity, science knowledge, and engineering design process.	Students design, create, and test a prototype of a wall (of specified criteria) to protect their school garden.

ferent objects as technology. Instead of sharpening a pencil, a child might be sharpening a piece of technology. The notion that everyday objects were technology had caused students to begin looking at materials in new ways, and we knew we could move onto the rest of the unit with confidence.

Lesson 1: Introducing Literacy and Science Concepts

Before every lesson, we shared science and language objectives with students, raising their consciousness of what we hoped they would gain from the lesson. For example, when the class read about a young girl from China and her efforts to build a garden wall to prevent a marauding bunny from eating her school's garden, we let the students know that we were building their readiness for an upcoming challenge. Our other objectives were to develop students' understandings of unit-related vocabulary; to help students identify the engineering problems and solutions embedded in the story; and to identify the materials used to create the designs in the story.

Before reading the book to our classes, we selected vocabulary from *Yi Min's Great Wall* (Engineering is Elementary Team 2005) and posted the words on the class word wall, along with illustrations and other pictures. For example, we showed pictures of the Great Wall of China, which was the setting for the story. One student, whose father was from China, became so excited by the story that she brought in Chinese toys and artifacts to share more about the culture of the story with her classmates! She was more excited about this unit of study than any other in the school year.

We also provided *realia*: real objects that students could see and touch to understand words and concepts. For example, instead of trying to define *clay* with words, we allowed students to manipulate a chunk of clay borrowed from the art room.

Most ELL students could easily access the word in their first languages; those with no experience with the material could now form a knowledge base. This was also science learning, as students would require firsthand familiarity with clay's properties for use in their designs at the end of the unit.

Later, when we read the story aloud, we reexamined these words in context. This allowed us to reiterate meanings and point out proper usage in Standard English sentences. Exploring language in context helped students make sense of the science and engineering story line.

Lesson 2: Materials and Their Uses

The second lesson, conducted during the following day's science period, was aimed at helping students understand the importance of using appropriate materials in technology. An engineer needs to understand the properties of materials in order to design objects that will meet the particular needs of a situation.

In this 45-minute activity, students were given a selection of different materials and were asked to choose which material would be the most appropriate for performing certain tasks. For example, one group was asked to find the best way to carry textbooks using cloth, string, or wood. They needed to describe how the material could be used appropriately to complete the task (the cloth could be sewn into a backpack, the wood could be made into a small wagon) in the most productive way.

Other groups' tasks included transporting eggs, keeping warm, and cleaning a floor. This activity helped students identify useful properties of materials used in technologies—a key aspect of materials engineering and a skill they would apply to using earth materials in the design challenge.

Lesson 3: Describing Earth Materials

Next, it was time for a two-session exploration of the science concepts related specifically to the design challenge. Students explored, compared, and described properties of different earth materials: first as dry samples, then mixed with water, and then mixed with each other and water. However, because both of our classrooms included students who need language support, we began the lesson by challenging students to describe the properties of materials that they encountered in their daily lives before moving on to the less familiar earth material mixtures.

To begin, we discussed the meaning of *properties* and synonyms used in other content areas. We immediately connected this conversation to hands-on experiences with materials. For example, while handling different samples, the second-grade students collectively generated adjectives to describe everyday materials—such as pieces of fabric or a piece of tape. We then created and posted an illustrated word bank with picture clues next to the words for the various properties.

We extended student understanding of the words and the materials by applying the words in contexts related to the design challenge. For example, to understand *sticky*, students touched pieces of transparent tape and applied them to different surfaces. Later, when students were working with clay, sand, straw, and wet and dry mixtures of these, we asked "Were any of these materials or mixtures sticky?" "How might stickiness be useful in walls?" These interplays of concept, language, words, and experiences

supported both science and literacy learning and prepared students to use this knowledge in a design context; later they would determine that a sticky substance is good for the mortar used to join bricks in a wall.

This time was well spent. Students showed surprise at the textures that resulted from mixing different earth materials with water. The silky smoothness of silt, the comparative rough or smooth qualities of differently sized gravel samples—these are materials with which our students had little experience. Taking the time to help them carefully observe the properties helped learners build an understanding of the diversity of earth materials and the vocabulary related to properties.

With the vocabulary and conceptual understandings firmly in place, we focused on scientifically exploring the potential building materials for the walls. We asked students to draw from their observations of the wet material mixtures and make (and later test) predictions about what properties these mixtures might have after they were formed into blocks and dried. The student teams shared results by completing a class chart—an additional opportunity for communication.

The combination of experience with the materials with an enhanced vocabulary and understanding of properties allowed students to approach the design challenge with an understanding of the engineering design process—asking pertinent questions, brainstorming ideas, planning, creating, and improving.

Lesson 4: The Design Challenge

We had now built some important scientific understandings of properties and had done so scientifically (systematically making mixtures and reporting their properties). Students were ready to approach the design challenge: Create a prototype of a wall to protect a school garden.

The second-grade class was presented with an imaginary scenario: The teacher had a garden that neighborhood children have accidentally been trampling, and thus the teacher needs to build a small wall around the garden to protect it. Using a doll to represent the trampling children, the second graders could envision the scaled-down prototypes of the walls.

The fourth-grade class drew from its own, true-life misfortune to complete the scenario. Wild rabbits from the woods surrounding our school had eaten several of our garden plants, which had been planted earlier to attract birds and butterflies. In both classes, students understood the need for a wall and could relate to the storybook character's need for one.

We established specific criteria for our walls: at least 45 cm long, at least 20 cm tall, and able to withstand the impact of a baseball rolled against the side. (Such criteria were important to emphasize the nature of engineering design as distinct from arts and crafts projects, in which children might build a representation of a wall that might or might not function to meet a specific need.)

Students received rulers and baseballs so that they would know exactly how to test their designs. Each class spent one 45-minute session individually brainstorming solutions and working in groups of four or five students to decide which design to use. They were to choose materials that they had worked with in Lesson 3 to create their designs; therefore, they discussed their results from Lesson 3 to come up with the design they thought would work best.

Students wrote synopses of their progress and designs and practiced verbalizing their thoughts before presenting them. During the design process, students were encouraged to use "think-pair-share," where they develop their ideas individually and discuss them with a partner before presenting to a group—thereby building verbal confidence. They also drew their ideas and organized and shared data. They connected their own experiences to the engineering design process. In short, students learned through a structured combination of experiences and expressions.

Students' designs included bricks of clay and straw mixtures, mortar with clay, and rock-embedded mud. All the while, children discussed, wrote about, and drew what was going on.

After all students or teams had created their designs, the class tested them with the balls and rulers. Once again, students engaged in scientific skills—conducting and interpreting "fair tests"—within an engineering context. Each wall was carefully measured to see if it met the required length and height. Then the baseball was rolled against it to see if the wall could withstand the impact. Then each student or team had a chance to improve the design, based on the wall's performance. This step provided students with important insights into the nature of the engineering design process and provided a critical opportunity to reconsider and refine the science ideas upon which they based their ideas.

Assessment and Evaluation

Given the linguistic challenges some of our students face, we opted against paper-and-pencil tests to assess science understandings. Instead, we used rubrics to implement embedded assessments of students' understanding of the engineering design process. Our rubrics—modified EiE templates—defined leveled evidence of learning in areas such as identifying materials, describing properties, and making predictions about and connections to relevant concepts. We also tracked individual and group progress, comparing it to our objectives.

Additionally, we evaluated students' drawings, diagrams, and final prototypes for evidence of understanding of the whole unit. Finally, we informally assessed individuals' participation, process, and language acquisition.

In the future, we might also compare successive prototypes. Design changes would indicate to what extent students arrive at and use new insights, understand and incorporate test results, and approach their processes systematically.

Reflections

Through the engineering design challenge, our students moved from a simple hands-on *building* experience into content by *designing* the walls. We asked students to think carefully about their walls and why some might perform well. We focused on properties students could identify that would be important for wall materials. Students helped create and define a list of adjectives to describe the materials we would explore and their results.

We observed that student involvement in engineering design units led to their learning valuable and transferable problem-solving skills as well as deep acquisition of science concepts. We found that students expressed their knowledge of science better and more creatively than we had seen when we taught from other hands-on science programs.

The activities allowed students to gain and demonstrate science understandings and language acquisition in multiple ways. We anticipate incorporating more design challenges into other science units, thus meeting some of our greatest challenges.

Resources

Engineering is Elementary Team. 2005. *Yi Min's Great Wall.* Boston: Museum of Science.

National Research Council (NRC). 1996. *National science education standards.* Washington, DC: National Academies Press.

Internet Resource

Engineering is Elementary
www.mos.org/eie

CHAPTER 15

Save the Penguins

Teaching the Science of Heat Transfer Through Engineering Design

By Christine Schnittka, Randy Bell, and Larry Richards ///

Many scientists agree that the Earth is warming and that human activities have exacerbated the problem (NRC 2001; 2002). Engineers, scientists, and environmental groups around the globe are hard at work finding solutions to mitigate or halt global warming. One major goal of the curriculum described here, Save the Penguins, is to help students recognize that what we do at home can affect how penguins fare in the Southern Hemisphere. The energy we use to heat and cool our houses comes from power plants, most of which use fossil fuels. The burning of fossil fuels has been linked to increased levels of carbon dioxide in the atmosphere, which in turn has been linked to increases in global temperature. This change in temperature has widespread effects on Earth, including effects on the life of penguins. If homes were better insulated, they would require less energy for heating and cooling, reducing fossil fuel use and carbon dioxide emissions. This is the problem presented to students in the beginning of the Save the Penguins curriculum.

In the Save the Penguins curriculum, students learn how engineers are addressing global warming by designing energy-efficient building

materials. Students learn the science of heat transfer; design experiments to test materials; and then assume the role of engineer to design, create, and test their own energy-efficient dwellings.

Penguins

Penguins are in peril. As the Earth warms, the oceans warm, pack ice melts, and penguins lose habitat. They also lose food sources such as krill, which rely on the protection of pack ice and feed on the algae that grow underneath the ice (Gross 2005). The emperor penguins in Antarctica are in severe decline due to climate change (Jenouvrier et al. 2008). South African penguins are actually leaving their nests to cool off in the water, placing their eggs at risk to attacks by gulls. Park rangers at Boulders Beach in Cape Town, South Africa, have created little semi-enclosed "huts" for penguins to nest in, which keep them cooler and protect their eggs from predation (Nullis 2009). Several short videos found on YouTube help engage students in this issue and address the impacts of global warming (see Resources). If your school blocks access to YouTube, download and save the videos at home using an online tool such as the one provided free at Downloader9. com (*www.downloader9.com*), then save them to a disk or flash drive for use at school.

Setting the Stage for Engineering Design

The problem presented to students in this curriculum is an analogy with symbolic meaning—to build dwellings that protect penguin-shaped ice cubes from increasingly warming temperatures. Through their work on the project, students learn about thermal energy transfer by radiation, convection, and conduction. Students test materials for their ability to slow thermal energy transfer in order to keep the ice penguins cool. After testing

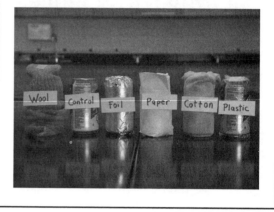

FIGURE 15.1.

The cans demonstration

materials, students build their penguin homes, and then see how well the dwellings keep the penguin-shaped ice cubes from melting in a test oven.

Students' Alternative Conceptions

Involving students in a series of discrepant-event demonstrations helps them form more scientific understandings of heat transfer. The five demonstrations that follow can be used to target common student misconceptions. Other demonstrations are possible, but the following were used in research and empirically shown to be effective (Schnittka 2009).

The Cans: Understanding Insulation and Conduction

Materials

- six cans of soda refrigerated overnight
- paper towel

- aluminum foil

- plastic wrap

- wool sock

- cotton sock

- six thermometers

If you ask students what they would wrap around a can of soda to keep it cold on a field trip, most will suggest wrapping it in aluminum foil. They might explain that aluminum foil "keeps the cold in." They would not dream of wrapping the can in a wooly sock, due to the prevalent conception that wool makes things warm. Demonstrate this fallacy by wrapping cold cans in different materials and taking their internal temperatures several hours later. Be sure to include a control (no wrapping) for fair comparison (see Figure 15.1). Have students make predictions about which can is the coldest. They are usually quite surprised to find that the wool sock keeps the soda coldest. Use this demonstration to help your students understand what insulators and conductors are. This demonstration and the interpretive discussion that follows usually take 20 minutes.

The Trays and the Spoons: Understanding Why Metals Feel Cold

Materials

- plastic tray

- silver (or silver-plate) tray

- thermometer strips

- silver (or silver-plate) spoons

- plastic spoons

- penguin-shaped ice cubes

FIGURE 15.2.

Plastic and silver-plate trays

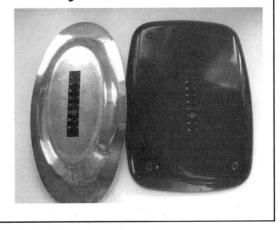

FIGURE 15.3.

Taking the temperature of the trays

To prepare for the following demonstrations, borrow or bring from home two trays—one metal and one plastic (see Figure 15.2). Tape an aquarium thermometer strip to the underside of each tray. Each thermometer should display room temperature (see Figure 15.3). Flat LCD thermometers in a large, easy-to-read size can be purchased from PetSmart (*www.petsmart.com*). Alternatively, Ideal brand #61-310 multimeters with thermocouple probes can be used. They can be purchased from a number of online sources.

For the first part of the demonstration, walk around with the trays and allow students to touch them and describe which tray feels colder. While both trays are at room temperature, students will insist the silver tray is colder. After the demonstration with spoons (description follows), reveal the actual temperatures of the

trays to students. Students should understand that both of the trays are the same temperature, but that heat transfers faster from a warm hand to the silver tray than it does from a warm hand to the plastic tray.

The second part of this demonstration involves spoons and penguin-shaped ice cubes. Lékué brand penguin ice cube trays (model # 39004) can be purchased online from a number of sources (see Figure 15.4). Give each student group a plastic and a silver spoon to hold. Silver plate or stainless steel will suffice, but silver is a better conductor of heat. Students may comment that the silver spoon feels colder and infer that metals are naturally colder than plastics, just as they may have done with the two trays. Have students predict which spoon works best to keep an ice cube from melting. Most students predict the silver spoon because it feels colder, as did the silver tray. Place penguin-shaped ice cubes on the spoons and have students take

FIGURE 15.4.

Lékué brand penguin ice cube tray

FIGURE 15.5.

Silver and plastic spoons with penguin-shaped ice cubes

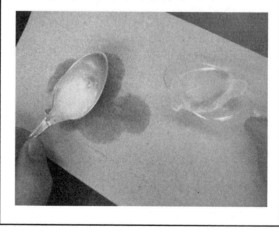

turns holding the spoons (see Figure 15.5). Ask students to explain why the penguin in the plastic spoon does not melt significantly, while the penguin in the silver spoon quickly turns to water.

Through group and class discussions, bring students to an understanding that heat transfers from where the temperature is higher to where the temperature is lower, and that heat transfers faster through the silver spoon than the plastic spoon because silver is a better conductor than plastic. The silver spoon feels colder because the hand, which is warmer, is losing thermal energy in the heat transfer. While students may try and insist that cold transfers because they feel their hands get cold, remind students that there is no such thing as cold transfer; only thermal energy, or heat, can transfer. The feeling that the hand is cold when touching the metal is evidence that the hand is a very good detector of rapid heat loss.

Next, reveal the equal temperatures of the silver and plastic trays. This discrepant event promotes a lively discussion and helps students come to a deeper conceptual understanding about the conduction of heat.

The House: Understanding Convection in Air and Radiation From Light

Materials

- cardboard house made from a box
- digital thermometers or thermocouple probes
- stopwatch or watch with second hand
- computer with spreadsheet software (optional)

The following demonstrations help students understand that convection occurs when less-dense fluids rise and more-dense fluids fall. It can also be used to demonstrate how black surfaces absorb radiation and become quite hot, while reflective surfaces do not. These demonstrations and the interpretive discussions that follow usually take 20 to 30 minutes.

Make a hollow cardboard house and paint the roof black with acrylic paint. Shine a shop light onto the house, and place temperature probes in the attic and the first floor. The

FIGURE 15.6.

Data and graph from house demonstration

Time (sec.)	Attic temperature (°C)	First-floor temperature (°C)
5	37	22
10	36	23
15	35	24
20	34	25
25	33	26
30	32	27
35	31	28
40	30	29
45	29	28
50	28	27
55	27	26
60	26	25

attic gets quite hot inside, approaching 38°C (100.4°F). (**Safety note:** The cardboard black roof gets very hot and can start to smoke if the shop light is placed too closely. Be sure to monitor this, and have a fire extinguisher nearby just in case.)

After you turn off the light, ask students why the attic remains hotter than the first floor. The likely answer is "Heat rises!" The scientifically correct explanation would be that cooler air falls because it is more dense (particles are closer together), and this forces hot air, which is less dense, to rise. There are several popular demonstrations of convection involving water and food coloring that can be used, but this demonstration helps students understand that convection is the movement of all warmer and cooler fluids (liquids and gases), not just water.

After the light has been off for a moment, flip the house upside down. The air masses change places due to density differences, and the thermometers or temperature probes reveal this change. Have students record the temperature of the attic and first floor as the air masses change places. They can graph the data by hand or enter the data into a spreadsheet to graph the changes over time. Without access to computer-based probes, we usually have one pair of volunteers, wearing safety goggles, call out the temperatures while one volunteer calls out the "time" every five seconds and another pair of volunteers record the temperatures. See Figure 15.6 (p. 113) for sample data collected from a demonstration.

Make sure students come to understand that the air masses are changing places as cooler, more dense air falls and warmer, less dense air rises. You might also want to repeat the demonstration with several layers of aluminized Mylar film protecting the house, as

seen in Figure 15.7, in order to demonstrate the difference between a black roof and a reflective roof (optional).

The Shiny Mylar: Understanding Radiation and How to Reflect It

Materials

- shop lights (one per group with 150 W light bulb)
- ring stands (one per group)
- several sheets of aluminized Mylar (per group)

FIGURE 15.7.

Mylar on a hot black roof

The following demonstration helps students understand that light can be reflected, preventing heat transfer from radiation. This demonstration and the interpretive discussion that follows usually take 15 minutes.

Mylar is a polyester film that can be coated with aluminum to be very shiny. It is usually thin and can be somewhat translucent, but when layered, it is excellent at reflecting light and preventing heat transfer. It is commonly used in snack-food packaging, helium-filled balloons, and emergency blankets. We have found that the easiest and least expensive way to purchase aluminized Mylar is at a craft or stationary store as metallic tissue wrapping paper.

Have students feel the radiation from the lights with their hands. (**Safety note:** Make sure students keep a safe distance from the shop light. To do this safely, have the shop light mounted on a ring stand at a safe distance pointing toward the counter.) Have students working in groups take turns placing their hands on the counter under the lamp as their partners slip the shiny Mylar layers in between the light and their hands. Students will immediately feel the reduction of heat transfer, especially with multiple layers. Use this demonstration to help students understand that radiation is a type of heat transfer that occurs nearly instantaneously, at the speed of light, and that the light is easily reflected by shiny materials. Students are amazed at how suddenly the heat transfer is blocked the moment the Mylar is placed between their hands and the light.

Creating a Dwelling for Ice Cube Penguins

Testing Building Materials

After students have a good, basic understanding of how heat transfers, introduce the challenge:

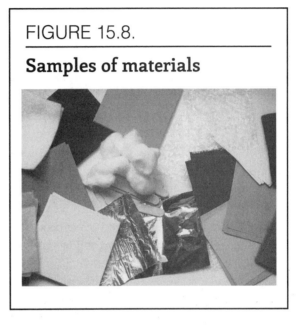

FIGURE 15.8.

Samples of materials

to build a dwelling for a penguin-shaped ice cube in order to keep the penguin from melting. First, students must decide which materials to build with.

Students engage in scientific inquiry as they test materials and eventually design and test dwellings. Student groups are given materials such as felt, foam, cotton balls, paper, shiny Mylar, and aluminum foil to test for their effectiveness at preventing some form of heat transfer (Figure 15.8). It is best for the teacher to prepare all of the samples ahead of time. We find that cutting the fabrics, foils, papers, and foams into 7.6 × 7.6 cm (3 in. × 3 in.) squares with a 7.6 × 45.7 cm (3 in. × 18 in.) clear acrylic quilting ruler on a cutting mat with a rotary cutter works best. These items can be purchased from sewing stores. (**Safety note:** All safety protocols need to be followed with building and cutting—students should wear safety glasses.) Decide as a class which materials or combinations of materials can be compared and tested, then divide the work up among the different lab groups. It is reasonable for each group to run three or four tests.

For example, one group might compare aluminum foil to shiny Mylar, while another group compares white felt to white foam. Some groups could test bubble wrap with different colored papers on top. Students can compare materials under a shop light mounted to a ring stand shining on a black surface such as a black countertop or black plastic tray. Give students access to thermometers and timers to fairly test samples under the light or on the hot black surface. As students explore the materials, they begin to formulate ideas about how to build a dwelling for a penguin-shaped ice cube so that the least amount of ice is melted. Allow students to procure additional materials after they decide which ones are better building materials. We usually price the materials and give each student group a budget to work with.

Encourage open inquiry and allow students to come up with their own testing ideas. Be sure student groups share the results of their experi-

ments so that the knowledge gained is communal, encouraging a more collaborative and less competitive environment. During sharing time, it is helpful to encourage students to explain to each other why they tested certain materials, what their results were, and why they think the results turned out as they did. This is the perfect opportunity to help students understand heat transfer as it applies to each experiment performed. Expect students to spend at least one class period testing materials and half of a class period sharing their results with each other.

Building the Dwelling

After students have tested different building materials and shared their results with the class, they will be eager to start using the materials to build little houses to keep the penguin-shaped ice cubes from melting. Students will finally be able to apply what they have learned from the demonstrations, discussions, and materials testing. As students take on the role of "engineer" to keep ice from melting, remind them that engineers are designing innovative materials for houses, schools, and other buildings to prevent heat transfer. Preventing heat transfer is energy efficient, and efficient buildings use less energy. The less energy needed to heat and cool a building, the less negative impact it has on the environment.

Provide students with tape and glue, and scissors to cut paper and fabrics. Most students will use the materials as given, but some will want to modify them by cutting, folding, crumpling, or layering. Students need to be sure to create an opening so that the penguin can be easily placed in the dwelling and easily removed after testing. As the designs evolve, students may need to conduct further testing, discuss results with other groups, and receive support for their ideas from the teacher. A class period should be sufficient for this phase of the curriculum. Engineers work

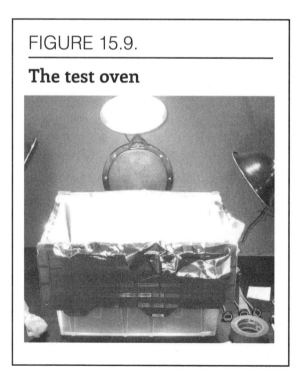

FIGURE 15.9.

The test oven

FIGURE 15.10.

Little dwellings in the oven

with constraints of time, space, and money, so give your students a time deadline, too.

Testing the Design

Eventually, students are ready to evaluate their dwellings in the oven. The oven can be a large plastic storage bin lined with aluminum foil on four sides and spray-painted black on the bottom, with three 150 W shop lights shining inside so that all three forms of heat transfer can occur. If you use a black plastic storage bin, simply line the sides with foil and attach the shop lights. See Figure 15.9 for a suggested test-oven design. Houses placed in this preheated oven experience conduction with and radiation from the black floor, radiation from all sides, and convection as hot air rises off the black bottom.

Fill the ice-cube trays the night before with 10 ml of water for each penguin. We use a medical syringe to be accurate. The carefully created and frozen 10 g ice penguins are simultaneously placed inside the dwellings and then placed in the oven and subjected to 20 minutes of intense heat. See Figure 15.10 to see how dwellings are placed

in the oven. The oven can accommodate more dwellings depending on its size, but students need to be told about the space limitation. Typically, we have eight teams in each class, so we tell students that their dwelling cannot be larger than one-eighth of the floor space in the oven.

During the 20-minute wait, we sometimes show students a slide show on energy-efficient building materials such as smart windows, aerogels, radiant barriers, and solar-panel roofing tiles. Sometimes we show an excerpt from the movie *March of the Penguins* or an excerpt from *An Inconvenient Truth* to spark discussion. After 20 minutes, the dwellings are finally removed and the remaining solid portions of the ice penguins are placed in little plastic cups for mass measuring. (Determine the mass of the plastic cups ahead of time and write the mass on the cups.) Students always enjoy the thrill of rescuing their dwelling from the oven and seeing how much penguin they were able to save. After the solid penguin remains are placed in plastic cups, it will not matter if some melting occurs prior to measuring the mass because the water stays in the cups. This provides a teachable moment to discuss melting and the conservation of mass. It usually takes a few minutes for each group to take a turn determining the mass of their cup and penguin on a digital mass scale, then subtracting the mass of the cup. (**Safety note:** Mount the test-oven shop lights on ring stands so that they can be turned off and moved away when students place their penguin dwellings in the oven or retrieve them.)

The results are shared and discussed after all penguin-ice-cube masses are measured. Some student groups are able to save at least half their penguin, but some usually only have a few grams remaining.

It is always interesting to discuss which design features were best at preventing conduction with the black oven bottom. Which design features were best at preventing radiation from the heat lamp from penetrating the dwellings? Which design features were best at preventing the convection of hot air moving? Students discuss and decide. They analyze the results, and then, if time is available, it is back to the drawing board to make revisions and improvements for a second round of testing.

When time is available, students are able to use their shared results and ideas to make revisions that help save even more ice penguins. This step helps to mitigate competition, because each group of students is a winner if their revised design is better than their first one. The process mimics how engineers continually work together in an iterative process to make things better. It usually takes students an entire class period to redesign their dwellings, bake them in the oven, measure the

mass of the penguins, and discuss the final results.

See Figure 15.11 for an initial and revised design after students were able to learn from their first experience and share results with each other. Notice that students decided to plug up the door to prevent convection and add insulation in the "tube" to prevent the ice from directly contacting the metal. Many students decide to put their dwelling up on stilts of some sort, and many provide extra protection from radiation on all sides.

When the second trial is conducted, or when results are compared among classes, make sure the ice is at the same starting temperature each time. Ice will be the temperature of the freezer when it is removed, and it warms until it becomes 0°C (32°F) and melts. The standard freezer temperature is –17°C (1.4°F), but each freezer can vary. Therefore, the results will be the most accurate if

FIGURE 15.11.

Initial and revised designs

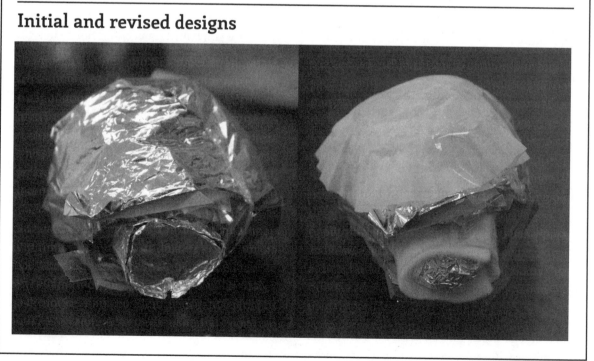

penguins are removed from the same freezer each time they are used, and used right away.

Conclusion

This unique engineering, design-based approach to learning science in a context relevant to students' lives has been shown to be effective in (1) improving students' conceptual understandings about science, (2) increasing their knowledge about and attitudes toward engineering, and (3) increasing their motivation to learn science (Schnittka 2009; Schnittka et al. 2010). Research using the Save the Penguins curriculum has shown that with scaffolding, students can connect the heating and cooling of their homes to the burning of fossil fuels for energy production, global warming, melting sea ice, and penguins in peril. In the meantime, students are sharpening their inquiry and process skills, working in teams, and thinking creatively. Students learn the science best and engineer better dwellings when the demonstrations are used early on to target their alternative conceptions about heat. Engineering design activities can be used effectively in science classrooms, but careful attention should be paid to pedagogically sound science teaching. Students' alternative conceptions must be acknowledged, addressed, and targeted in science lessons. Otherwise, there is no expectation that students will implicitly learn scientific concepts just because they participate in an engineering activity. The engineering design activity gives students a tangible application for their science conceptions, insight into the world of engineering, and practice with 21st-century skills: innovation, problem solving, critical thinking, communication, and collaboration.

The complete Save the Penguins curriculum with daily lesson plans is available online at *www.uky.edu/~csc222/ETK/SaveThePenguinsETK.pdf.*

References

Gross, L. 2005. As the Antarctic ice pack recedes, a fragile ecosystem hangs in the balance. *PloS Biology* 3 (4): 557–61. *www.plosbiology.org/article/info%3Adoi%2F10.1371%2Fjournal.pbio.0030127.*

Jenouvrier, S., H. Caswell, C. Barbraud, M. Holland, J. Stroeve, and H. Weimerskirch. 2008. Demographic models and IPCC climate projections predict the decline of an emperor penguin population. *Proceedings of the National Academy of Sciences* 106 (6): 1844–47. *www.pnas.org/content/106/6/1844.full.pdf+html.*

National Research Council (NRC). 2001. *Climate change science: An analysis of some key questions.* Washington, DC: National Academies Press.

National Research Council (NRC). 2002. *Abrupt climate change: Inevitable surprises.* Washington, DC: National Academies Press.

Nullis, C. 2009. *Save South Africa's penguins: Give them a home.* Associated Press. March 29. *http://addistimes.com/africa-news/1873-save-south-africas-penguins-give-them-a-home.html*

Schnittka, C. G. 2009. Engineering design activities and conceptual change in middle school science. PhD diss., University of Virginia.

Schnittka, C. G., M. A. Evans, B. Jones, and C. Brandt. 2010. Studio STEM: Networked engineering projects in energy for middle school girls and boys. Proceedings of the American Society of Engineering Education, Louisville, Kentucky. *http://soa.asee.org/paper/conference/paper-view.cfm?id=23015.*

Internet Resources

Climate Change Likely to Devastate Emperor Penguin Populations in Antarctica
www.youtube.com/watch?v=RqNJ6B1CSss
Penguin in a Pickle
www.youtube.com/watch?v=Jz-5Y7WgVEE
Penguins Are Melting
www.youtube.com/watch?v=rqUvf9Rxxj4

CHAPTER 16

Shake, Rattle, and Hopefully Not Fall

Students Explore Earthquakes and Building Design

By Adam Maltese

Earthquakes occur across the globe, and their effects can be felt by people regardless of location. However, a moderate earthquake in Pakistan or Turkey may cause much greater damage than a stronger earthquake in Tokyo. It is imperative to help students understand why this disparity exists—often due to both natural and human influences. Students often ask, "Why don't all the tall buildings fall down when there are earthquakes?" Through this activity, my sixth-grade students began to understand the engineering challenges of building earthquake-resistant buildings and how scientists meet that challenge.

The goal of this project, which I conducted for four years while a sixth-grade teacher in Greenwich, Connecticut, was for each group of students to erect a model building that could withstand shaking from simulated earthquakes. Students were to use toothpicks and glue for the main structure of the building but were free to use materials of their own choice to design and construct a *base isolation system*, which separates a building from the Earth and reduces the amount of energy transferred to the building during an earthquake (see Figure 16.1, p. 122, for more information on this topic). By using a hands-on activity to investigate earthquakes, this project engaged students with the material, covered the appropriate content standards for Earth science and natural hazards, and gave students an outlet to demonstrate creativity in the design of their buildings and base isolation systems.

Project and Proposal

After a few lessons covering general background about earthquakes, I introduced the building project. As a result, students came to the project with some knowledge of why

---

FIGURE 16.1.

Base isolation systems

Skyscrapers are usually built with a steel frame attached to a foundation, and foundations are anchored to the bedrock or geological material that sits beneath the building. When an earthquake occurs, the energy from the moving earth will be transferred into the structure through its foundation, creating the potential for significant structural damage. Therefore, if it is possible to separate a structure from the shaking earth, this can reduce the amount of energy transferred to the building during a quake.

As the name implies, *base isolation systems* are engineered to divorce a building from the foundation and earth on which the building sits. These systems are meant to reduce the amount of acceleration that a structure might endure during an earthquake, which, in turn, reduces the likelihood of severe damage. Isolation systems include rubber dampers, sliding bearings, and, more recently, magneto-reactive fluids (see Internet Resources for more information on earthquake engineering methods).

Rubber damper

Sliding rail system

- **Rubber damper**: Rubber plates placed between the building and foundation reduce motion and forces absorbed by the building.

- **Sliding rail system:** Such systems allow buildings to slide in reaction to moving earth below.

- **Magneto-rheological fluids:** Fluids, usually made of iron particles suspended in oil, become rigid when a magnetic/electric field is induced around the suspension and can dampen the transfer of motion from ground to building.

---

earthquakes occur, the motions of the Earth's crust during a quake, and the associated hazards in different geologic settings. The work and assignments explained below were spread out over the next three to four weeks of class (45-minute periods, 4–6 times) while instruction on earthquakes and other natural hazards continued.

The project began with a proposal stage. After the class viewed videos that depicted the devastation earthquakes can cause and the varying degrees of damage (based on population density, predominant structural materials, and building codes), I provided an initial explanation of the purpose of base isolation systems and then introduced the project: "Your goal is to build a structure that can withstand an earthquake by using a system that isolates the building from the earth on which it sits."

For the isolation system, the students were urged to come up with creative designs that perform the necessary task—isolating the building from the shaking ground. Students were exposed to designs that engineers currently use in the videos, but my expectation was that they'd try something "new" to accomplish the task.

Students formed groups of 2–3 and were given some time to think about their designs. During this time, I made myself available for consultation but never provided any concrete design ideas. Over the next week, the students were given a few spare minutes of class time and allotted time (as homework) to prepare a one-page proposal of their plans for the structure and the isolation system.

An important part of the proposal was the materials list. Reviewing this list gave me a sense of how elaborate each design was and allowed me to cut off potential design disasters before construction began. If necessary, I provided feedback on how students might recraft their plans, basing my ideas on fundamental physical science principles. For example, designs that fought against gravity (e.g., upside-down pyramids) or heavy structures set on bases without adequate structural support were recommended for redesign.

Let's Get Building!

Once the designs were approved, construction began. I outlined the following requirements to students:

1. All buildings must stand at least 0.5 m above the base to limit materials. This height seemed to allow students to spend half their time on the base and half on the structure. Also, anything shorter was too stout for a high-rise and anything taller was too flimsy.

2. To establish a level of equality across the groups, students were required to build the main part of their structures using toothpicks and glue sticks (these are readily available, affordable, and generally consistent in quality). Each group was provided a maximum of five boxes of toothpicks and 12 glue sticks (for hot glue guns) to hold the structure together. (Before I implemented this regulation, a group of students built a solid tower of toothpicks and glue. Although that building was able to support weight, the structure lacked any usable space for potential occupants.) Bottles of glue were also made available but are generally more difficult to use because the bottled glue is less viscous. Most students have experience using glue guns in art class, but make sure to allow for a little testing time and give clear safety instructions for avoiding burns.

3. There were no restrictions on what students could use for the base. Students provided their own materials for that portion.

4. Another requirement was that the structures withstand the internal load of a masonry brick for 30 seconds. Real buildings must sustain not only the weight of their structures but substantial contents (e.g., people, computers, books) as well. Using the standard of a brick gives students an identifiable quantity to pretest the strength of their buildings.

5. The last requirement was that the buildings withstand "earthquakes" mimicked by our shake table (Figure 16.2). The structures are tested by securing them to a shake table, so the base cannot be larger than the shake table surface, 30 cm × 30 cm for my class.

To minimize any parental influences, students were given time during class (typically four or five 45-minute class sessions) to complete their structures. Many students quickly realized they could mass-produce sections of their building, thus creating an assembly line and improving efficiency. Having the majority of project work completed in class allowed me to supervise and ensure that all members of the groups were getting a chance to contribute.

FIGURE 16.2.

Shake table construction

This is a photo of a building base (note no isolation system used) attached to the shake table surface with small clamps. Hidden under the shake table platform are the four casters installed to support the sliding surface.

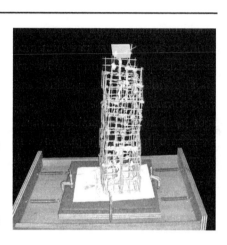

A search for "earthquake table" on the internet will yield a number of designs of varying complexity (see Internet Resources). Our design is not perfect but was straightforward to build, moderately priced, and durable. I began by building a shallow wooden box, approximately 1 m long by 0.5 m wide. Four spherical casters were then attached (wheel-up) near the center of the box bottom. A wooden plank was placed on top of the casters and attached to the sides of the box with springs and eye hooks, placing the springs under moderate tension. This wooden plank acted as the "bedrock" to which the building bases were fastened with clamps. One limitation of this design was that it allowed only for testing the horizontal motion component of an earthquake. The springs used were a bit on the tight side; I recommend buying a few different varieties (at local hardware store) to limit frustrating return trips during construction. Whatever type of table you build, test your design until you are satisfied with the quakes that are produced. We were able to put the table together for under $30, and it was durable enough to last a few years of testing.

Shaking and Testing

The groups' final grades were composed of a score for aesthetics, design, load-bearing support, and the ability to withstand simulated earthquakes (Figure 16.3). Before the testing of each building, students explained their design choices to classmates. Depending on the number of groups involved, testing usually took one or two class periods to complete.

First we did a load test without shaking. A brick (weighing a few pounds) was placed on each building top for 10 seconds, and this was repeated three times. Most buildings survive this, but a few suffer minor structural damage. After this simple test, we moved on to shaking.

The shake testing was done in two phases: unloaded and loaded. This allowed for buildings to receive partial credit for surviving the unloaded test yet failing the loaded test. To conduct the tests, the building base was fastened to the shake table with small clamps. To simulate an earthquake, the table was pulled to one side of the box, compressing one set of springs and stretching the opposing set. The table was released, causing the attached building to accelerate back and forth for approximately one second. Each test was conducted three times to average out variations in the induced earthquakes. All but the weakest structures usually survive the unloaded test.

For the loaded tests, a brick was placed on the structure before conducting each simulated earthquake (again, three trials). If the design of a building did not allow for the brick to be fastened to the structure with tape or Velcro, I asked a volunteer to "spot" the brick during the test to reduce the chance of injury from a falling brick. For the loaded tests, the failure rate was approximately 35%, with the majority of buildings withstanding the tremors.

Follow-Up

After the testing phase was completed, each student prepared a 1–2 page project summary. Students reflected on difficulties they experienced with the design and construction and assessed the performance of their isolation systems. I encouraged students to use various sources to find out how isolation systems are employed in real-world structures and to compare those systems with their base designs. Many students' systems employed mechanisms for damping (e.g., layers

---

FIGURE 16.3.

Sample grading breakdown for each group

Aesthetics (10 pts): Subjective measure of the appearance of the building structure and the surrounding landscape

Height (20 pts): To earn maximum points, structure has to meet minimum height requirements (0.5 m); 1 point subtracted for every centimeter below the required height

Supports One Brick (20 pts): Structure must withstand the load of one masonry brick (weighing 5 lbs) for 30 seconds

Base Isolation System (25 pts): Subjective measure of the creativity and logic in the design of the isolation system; any type of isolation system received at least 15 points

Withstands Earthquakes (25 pts): Structure must withstand three earthquakes with the structure unloaded and three earthquakes with the structure loaded to receive full points; partial credit was dependent on the number of tests survived by the structure

---

of jello and wood) or sliding (e.g., plates separated by ball bearings) that were similar in principle to systems employed in actual buildings.

Learning Outcomes

My main goal in assigning this project was to tie information students learn in class to a real-world application. Because I wanted students to have a lot of freedom in the project, I provided only minimum guidance—this approach did not always result in students making all the connections with the content that I hoped for. For example, some students built sturdy structures but neglected to make their buildings flexible enough to withstand shaking. Others designed systems that could compensate for one type of movement but not all three movements associated with big earthquakes. However, in the designs, presentations, and follow-up reports, it was clear that students understood how earthquake waves move through the Earth and that it is often the interaction of natural (e.g., earthquake intensity) and human (e.g., building codes) factors that influence the level of devastation caused by seismic events. Additionally, I felt this project increased student engagement in class. During the earthquake unit, students frequently asked how certain concepts would relate to improvement of their designs.

There are many opportunities to extend this project further into the fields of engineering and Earth science. For example, students could investigate different shapes and materials (e.g., pasta or straws) for building the structures. Or, students could be challenged to build the strongest structure or a building using the least materials. Finally, students could investigate data on earthquake magnitude and damage to assess the efficacy of building materials and methods used in different parts of the world.

Looking back at the finished buildings of a few hundred students, I am impressed with the creativity that students employed in all parts of the project. Whether it was a creative use of marshmallow-filled ziti for a base or a hexagonal building design, the students applied their knowledge of earthquakes with an intuitive knowledge of structural stability to tackle a real-world problem.

Internet Resources

Earthquake Information from the USGS
  http://earthquake.usgs.gov
Lord Corporation (Lord) Magneto-Rheological
  Technology
  www.lord.com/Home/MagnetoRheologicalMRFluid/
  tabid/3317/Default.aspx
Multidisciplinary Center for Earthquake Engineering
  Research
  *http://mceer.buffalo.edu/infoservice/reference_*
  services/advEQdesign.asp
Pacific Earthquake Engineering Research Center
  http://peer.berkeley.edu
The Learning Channel—Tremor Tech
  http://tlc.discovery.com/convergence/quakes/
  articles/tremortech.htm

PART TWO
Content Area Activities

PHYSICAL SCIENCE

CHAPTER 17

Gravity Racers

By Dawn Renee Wilcox, Shannon Roberts, and David Wilcox //

C hildren experience the principles of motion on a daily basis as they play with toy cars, zoom down a slide, or coast downhill on a sled. With the 2010 Winter Olympic Games prominent in the media, our children were exposed to images of athletes skiing down snow-covered slopes, coasting furiously on bobsleds, and skating gracefully across the ice.

We capitalized on our children's natural curiosity about the world around them by exploring the concept of motion. Our weeklong series of science, technology, engineering, and math (STEM) activities provided the opportunity for us to lead a 21st-century instructional approach to science education that involved hands-on and direct experiences for our students. To do this we used a 5E learning cycle model (Engage, Explore, Explain, Elaborate, Evaluate) as students investigated force and motion (Bybee 1997). Students designed, built, and tested their own simple gravity-powered car using experimentation and active investigation; essentially, they modeled thinking processes that are similar to those applied by scientists as they gathered information. We carried out this activity in a fourth-grade classroom, but it could be used with grades 3–5 with slight modifications.

Engage

The experience began with a challenge encouraging students to design a racing vehicle for a new Olympic racing event, "Gravity Racers." In this event, athletes would travel down a slope (inclined plane) in a straight line for a determined distance powered only by gravity. We told the students that they would work in design teams and would be in charge of conducting experiments to develop an understanding of the position of an object and of the forces that cause that object's motion.

Explore

The lesson continued with an exploration session. Students assembled into groups of three to five students. They were encouraged to explore using a ramp and a variety of objects. Our ramp was a simple sheet of thin wood propped against a 10-cm high stack of textbooks. The objects included vehicles with wheels, balls, blocks of wood, marbles, or pennies. We encouraged students to discover movement by allowing them to explore freely, test their predictions, and discuss their ideas about motion with others.

As the students became comfortable in their explorations, we encouraged them to move toward more formal data recording. Students were asked to create their own series of mini-experiments. The instructions to the students followed a simple format with each object. First, they asked themselves, "How will my object move?" Next, they predicted how their object

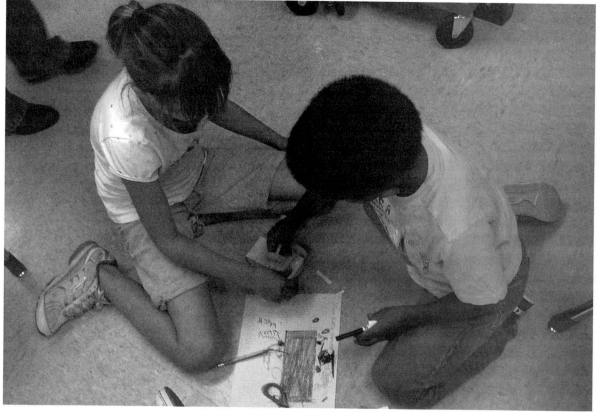

Students explored motion before designing their vehicles.

would move. Last, they investigated how their object moved. Students used a handout titled *How does it move?* to guide their investigations (see NSTA Connection). They recorded which objects moved in straight or curved paths and which vehicles traveled the farthest using simple data tables. They determined the average distance traveled by each vehicle. This data helped facilitate a discussion in an effort to increase student understanding during the explanation portion of the lesson.

During recess we took the students on a "motion walk" around the school yard, stopping at intervals to notice examples of motion. When we returned to the classroom, we created a list of examples. Those examples included balls dropping and bouncing, a teacher pushing a cart, wheels on the cart rolling, and students gliding down a sliding board. The engagement, exploration, and motion walk provided a foundation for later activities in which we took these common real-world (playground) experiences with motion and gradually introduced the concepts of forces, reactions, gravity, friction, and inertia.

Explain

Next, we allowed students to reveal what they already knew about force and motion and gave them the opportunity to organize and build on background knowledge. The goal was for students to realize that every movement is caused by a push or pull and to develop an understanding of the concepts of inertia and friction. To guide students, we asked them to analyze and classify their movements as pushes or pulls. We allowed students to discover on their own that a pull is sometimes gravity by creating their own explanations and listening to the explanations of others. Toward the start of the lesson, student answers indicated they believed that objects

stop on their own. One student commented, "The ball rolled down the ramp then stopped by itself." Another student remarked, "The car rolled fast, then it stopped. Yes, all by itself!" Later in the lesson student thinking shifted. Students realized that something must be acting on the object to stop it or slow it down. A student noted, "If you roll a ball down the ramp it speeds up because of gravity." We developed an understanding of *inertia*, the property of matter by which an object retains its state of rest or its velocity along a straight line so long as it is not acted upon by an external force. To encourage students to develop an understanding of inertia, we helped them build on prior knowledge and construct relations between force and motion. We demonstrated the old *pull the tablecloth out from under the objects trick*. The students saw firsthand that the objects on the table will not move unless acted on by an outside force. We also asked the students to experiment with stopping the objects and changing their direction. We asked questions like: Which objects roll? Why do some objects move without rolling? What causes the objects to slow down or stop? Initial comments included statements like, "The block won't move." We noted a conceptual change as student comments moved toward statements like, "If I push the block, it will slide down the ramp" and "The car hit Eleanor's leg, causing it to stop and change direction." These responses showed that students recognized that when things speed up or slow down, there is a cause.

We encouraged the integration of mathematics concepts by asking questions like: How far did each object move? Does the height of the ramp affect how the object moves? We carried the force and motion theme outside of our science time frame into our reading and language arts lessons. We encouraged students to explain their understanding of the concepts of motion through a vocabulary game and encouraged students to use the words as they carried out the challenge. A copy of the vocabulary matching game is presented in Figure 17.1. Students also read literature that related to the concepts (see NSTA Connection).

Elaborate

In the two or three days that followed, we gave students the opportunity to apply their new knowledge about motion as they participated in our motion design challenge. Students were required to collect and display data as they developed abilities to identify and state a problem, design/implement a solution, evaluate, and communicate with teachers and classmates.

Identify a Problem

We set the students up in design teams so that they could conduct experiments with motion and force to meet the design challenge. Our guidelines were that the car should have a chassis no smaller than 8 cm wide and no longer than 30 cm from bumper to bumper. We also explained that the vehicle should be able to roll down the ramp and travel for at least 100 cm.

Propose a Solution

Students were next asked to sketch a scale drawing—including measurements—of a vehicle to meet the design challenge. They communicated their design ideas to their teachers and their classmates in a show-and-tell fashion, allowing the opportunity for the teacher to capture student thinking.

FIGURE 17.1.

Vocabulary matching game

Energy	The ability to do work
Speed	A measurement of motion
Inertia	The property of matter that causes it to resist any change of its motion in either direction or speed
Friction	The resistance to motion created by two objects moving against each other
Kinetic energy	Energy in motion
Potential energy	Energy that is not in motion, but could be

Implementing a Proposed Solution

We gave students different materials to choose from (cardstock, construction paper, tissue paper, cardboard milk cartons, dowels, paper clips, pipe cleaners, straws, wheels, Styrofoam trays, spools, craft sticks, plastic bottles, glue, tape, scissors) as they followed the stages of technological design. The students worked collaboratively, using a variety of simple materials to create their vehicle. The materials listed are just a few examples of items that may be used. Feel free to use your imagination when choosing items for your students to use in their construction plan. In addition, rulers, stickers, paint, and other art materials might be useful to students.

Throughout the design process, we encouraged students to raise questions, develop hypotheses to test, and use their imaginations. We required students to record data in the forms of charts and diagrams and explained that we would use their individual or group data to create a class graph. We tied their data analysis scenario to the real world of

Olympic athletes as they measured and recorded their best times to determine who qualified to compete and win the medals. Students tested their cars by placing each at the top of the ramp and releasing them. Then we asked students to record the distance each team's vehicle traveled down the ramp and encouraged them to use that data to create graphs to show the distance in which their vehicles traveled. Students followed a scientific procedure to test their vehicle as they followed a prepared data-recording sheet (see NSTA Connection).

Evaluate and Communicate

Students analyzed their data and the data collected by other students to determine how well their design met the challenge. Students completed a checklist on their worksheet (see NSTA Connection). Again, students used measurements and communicated their findings orally and through writing. At this point in the lesson, we gathered the students together in one large group. Students reported their best times to the class and we recorded the data on our class chart. They modeled or talked about their successes and failures as they worked to meet the challenge. For example, one student commented, "I thought my car would go faster if I painted it orange. I learned that paint doesn't make it go faster." Students were able to do this because throughout the process we encouraged them to make careful observations and discuss what was happening to their vehicle. We monitored the students as they discussed, examined, critiqued, explored, argued, and struggled with the challenge. We encouraged the students to explain the concepts and definitions in their words and to justify and clarify their ideas. Student comments included statements like, "A push or a pull is a force" "A push can change the direction of my car" and "A push can make my car go farther." We helped them under-

Each racer was tested to determine the best time.

stand the science behind the concepts by asking questions: What happened to the car? Did it travel in a straight line or turn? How might you get it to turn left or right? How far did it travel? How might you get it to travel farther? How might you get it to stop sooner or travel faster?

Not all of the students' cars traveled in a straight line the first time they rolled them down the ramp. Some cars curved to the right or left. One car had wheels that would not turn; that car just sat on the ramp. Students made design modifications and adjustments to their vehicles based on their findings, allowing for a discussion about cause-and-effect relationships and for us to focus on these changing quantities called *variables*. One student commented, "Wheels made it easier to push my car. The car without wheels takes more force to move." Another student reported, "When we tested our car the first time, our car rolled down the ramp and turned to the left. We adjusted our car by modifying how the straw [axle] was taped onto the car body and then it rolled straight." We introduced the students to the BBC Science Clips: Forces and Motion during this point of the lesson (see Internet Resources). The activity allows students to conduct a simulated experiment. The students can manipulate a number of variables and then push their car down the ramp. They are able to change the amount of force, the height of the ramp, and the size of the vehicle.

Assessment

We used an investigation rubric to assess understanding and progress (see NSTA Connection). Students were also asked to present an oral and pictorial version of their design and solution to communicate with classmates and teachers. Our students prepared a simple show-and-tell presentation. We used an oral communication rubric to assess understanding as well (see NSTA Connection). Other possible presentation methods might include: discussions, written reports, and computer presentations.

Extend Ideas

We finished by reinforcing the concepts. We extended the lesson by providing opportunities for students to apply their new knowledge in real-world situations outside of the classroom (e.g., relay races in physical education class, playground games, video clips, internet activities, bus ride home). At this point in the lesson, we visited Edheads Crash Scene Investigation website (see Internet Resources). This website contains information that helps kids learn about different professions and career choices in science.

Encourage students as they apply their new definitions, explanations, and skills to new but similar situations. The Rader's Physics for Kids! Motion website contains definitions and pictures that the teacher or students can access to reinforce the motion concepts (see Internet Resources).

Extend lessons by helping students summarize the relationships between the variables in the lesson. Teaching motion through STEM gave students the opportunity to connect real-world situations like playground games and sports to instruction in the classroom. The design challenge unified science, technology, engineering, and mathematics by placing the learning into a context that provides meaning and hands-on learning experiences intended to encourage students to develop an understanding of position and of the forces that cause an object's motion.

NSTA Connection

Download a reading list, rubrics, and worksheets at *www.nsta.org/SC1003*.

Reference

Bybee, R.W. 1997. *Achieving scientific literacy: From purposes to practices*. Portsmouth, NH: Heinemann.

Internet Resources

BBC Science Clips: Forces and Motion
   www.bbc.co.uk/schools/scienceclips/ages/6_7/ forces_movement.shtml
Edheads Crash Scene Investigation
   www.edheads.org/activities/crash_scene
Rader's Physics for Kids! Motion
   www.physics4kids.com/files/motion_intro.html

CHAPTER 18

The Egg Racer

By Jeremy Brown and John Corbin ///

Teachers are familiar with dozens of high school physics activities involving eggs. In this egg-related activity, students design and construct "Egg Racers" to learn problem-solving, physics, and engineering skills.

About 15 years ago, we set out to culminate a unit on electromagnetism with a project that involved electric motors. Inspired by the Massachusetts Institute of Technology design contests, we divided students into teams, provided each team with a bag of parts (e.g., motors, connectors, battery clips, sticks, switches), and directed the teams to build a machine that would score goals by hitting targets inside a homemade hockey rink. Since that first in-school design contest, we have tried several different projects. The most challenging, competitive, and engaging project has been the "Egg Racer."

The Racer

The Egg Racer project requires students (individually or in teams of two) to construct a vehicle to carry a chicken egg a distance of 3 m toward a terminal barrier (the wall) as fast as possible and stop before colliding with the barrier. [**Note:** Science Olympiad has a similar competition called the "Scrambler," which uses a falling mass for drive mechanism energy.] For the Egg Racer project, students are supplied with two 9V batteries, two DC motors, two paint-stirring sticks (donated by a local building supply store) for the vehicle chassis, two old CDs (as wheels), two snap battery connectors, and a simple slide switch. Small DC motors and other supplies suitable for this project can be obtained from a variety of vendors. The voltage rating for the motors is not critical and the cost varies from $1 to $2 each.

Students may supply their own wheels, to be used in addition to the CDs, and other miscellaneous parts such as gears, supports, or parts from toys (remote-controlled cars or parts are not allowed). Students are allowed four to five classroom periods for construction and another two periods for testing the finished product. Figure 18.1 shows an example of an egg racer. Students are not allowed to take their racers home to work on; however, they may consult with other people and resources for ideas and moral support.

Vehicles are judged on the amount of time needed to cover the 3 m, the amount of deviation from a straight line path, smoothness of operation, and stopping distance from the barrier. Hitting the barrier or dropping the egg en-route disqualifies the car for that run. [**Note:** The function of the egg is only to add suspense and risk to the endeavor. Any object, or no object at all, would suffice.] As a quantitative assessment, a scoring rubric provides feedback on the effectiveness of each vehicle (Figure 18.2). The best run out of three trials is taken for the final score.

The Race Track

The track is laid out on the floor of the physics lab and is 4 floor tiles wide × 10 floor tiles long. Egg Racers must start from within a designated starting zone and move in as straight a line as possible toward the barrier at the end of the track. For added challenge, small bumps and obstacles are sometimes placed on the track. Students, usually working with a partner, are allowed three tries on the track. In recent years we allowed only one racer on the track at a time in order to minimize crashes, but we have also run two at a time in past years. A practice track is available for students to test their vehicles as often as necessary. Often racers work fine when practicing, but the added weight and placement of the attached egg can cause breakdowns and frustration; therefore, we allow students to practice with an "egg substitute" (a 100 g mass).

Physics and the Racer

This applied physics project is a valuable teaching situation for problem solving, engineering applications, and design. Collision with the barrier is not allowed; therefore, the physics that concerns an object changing momentum during a collision does not apply. So what physics concepts do apply? Students learn about the operation of simple electric motors and the wiring of a simple series or parallel circuits with a switch. In addition, students also test circuits for best energy transfer applications. Students also have to consider friction effects and, in the more advanced designs, torque and gear ratio applications. (While students can hook up light bulbs and batteries in series or parallel without too much difficulty, they have a bit more trouble generalizing to motors in series or parallel.)

The main objective for this project, however, has evolved into teaching the design process. Students are given an opportunity to be creative in physics class and are exposed to a different kind of problem solving than what they are used to. This new problem-solving paradigm is the engineering-design process. We begin the projects by walking students through the "Design Process:"

1. Define the problem

2. State the design specifications

FIGURE 18.1.

Example of a student egg racer (notice the egg at front of car)

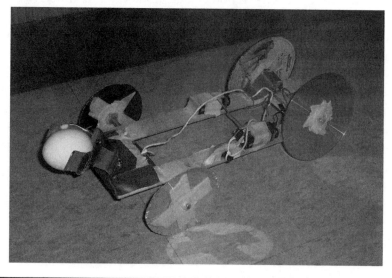

FIGURE 18.2.

Egg racer scoring rubric

Item	1	2	3	4	Your score
Egg holder	Simple taped connection	Egg is attached with rubber band; egg is firmly held in place	Egg is easy to insert and remove; egg is held firmly in place	Same as 3 with exceptional quality	_____
Speed	Vehicle moves	Moves within boundaries toward front wall	Straight and smooth forward motion, slight deviation	Straight, smooth, and fast with no sidewise deviation	_____
Stopping	Stopping mechanism included	Stops vehicle outside of 0.5 m range	Stops vehicle inside 0.5 m range	Stops vehicle inside 10 cm range	_____
Energy transfer	Poor power to drive mechanism, slippage, wobbles, intermittent drive	Little slippage of drive mechanism; slight wobbles	Smooth energy transfer, but slow motion of drive mechanism	Smooth energy transfer to drive mechanism	_____
Parts return				All recyclable parts returned in good shape	_____
				Total points	_____

3. Brainstorm

4. Identify design alternatives

5. Evaluate the alternatives

6. Test

After discussing the design process, we turn students loose and let them design racers.

Stopping the Racer

Getting the Egg Racer to move is not too difficult for students to accomplish; however, getting racers to stop at the proper distance is a different story. All stopping mechanisms must be onboard and deployed while the racer is in motion. No bumpers, sticks, or any other protuberances may extend beyond the front of the egg. We do not allow students to attach anything to the floor, so the "dog on a leash" approach is illegal. ("Dog on a leash" is when students anchor a string to the floor at the starting line so that the string will prevent the racer from crashing into the wall.) This rule effectively eliminates all stopping mechanisms except electric brakes, string brakes, screw brakes, or hybrid combinations of these.

Electric brakes have been designed in a variety of configurations, the most common of which is simply attaching one end of a string to the switch and the other end to one of the axles. When the racer has traveled the prescribed distance, the string is pulled taught; thus, tugging on the switch and turning off the power. String brakes are lengths of string attached between the front and drive axles. The string winds itself up on the front axle so that when the slack runs out, the string jams up the drive axle. Students quickly learn that they do not need 3 m of string to do the job because of wheel-to-axle ratios. Screw brakes are very elegant and precise. One of the axles is threaded so that a nut traveling along the threads jams tight when the racer has gone the correct dis-

tance. In all situations, rubber bands need to be affixed to the drive wheels to prevent skidding.

More advanced designs involve separate winding mechanisms to avoid binding up the wheels with excess string, or the use of gears or "belt drives" (rubber bands) to change ratios in case more torque or speed is needed. The most complicated braking system a student developed involved a motor rigged up as a generator, which in turn charged up a capacitor as the racer moved forward. When the capacitor became fully charged, it discharged through a transistor, which cut the power to the drive motor. The student tinkered with resistor and capacitor combinations to achieve the proper running time.

Future Skills

This activity aligns with the National Science Education Standards Science and Technology Standard (NRC 1996, p. 107). Several higher-order thinking skills are involved, including application, creative- and evaluative-thinking skills, and a host of problem-solving skills. The project usually occurs toward the end of the school year after exams, so students are ready for a change-of-pace activity and the project (which lasts five to seven classroom periods) does not interfere with traditional classroom work.

Each year we have had between 100 and 120 racers built. Virtually all students are able to complete the project on time; discover the design process; engage in a considerable amount of troubleshooting; and build something that operates to some degree of success. These skills are important and can contribute to future success no matter what field students choose.

Reference

National Research Council (NRC). 1996. *National science education standards.* Washington, DC: National Academies Press.

CHAPTER 19

Engineering for All

Strategies for Helping Students Succeed in the Design Process

By Pamela S. Lottero-Perdue, Sarah Lovelidge, and Erin Bowling   ///////////////////////////////////

As calls for science, technology, engineering, and mathematics (STEM) education at the elementary level become more vociferous, elementary teachers may be wondering whether engineering is meant for all students. They may question whether engineering is appropriate for their inclusive classrooms, where children with special needs are included in regular instruction.

We—an assistant professor of science education, an enrichment teacher, and a third-grade teacher—assert that engineering can be taught in inclusive environments. It may be especially empowering for those who struggle with traditional subjects. Here we describe how the core practice of engineering, the engineering design process, was taught in a third-grade inclusive classroom in which students used this process to design windmill blades. In this class, 7 of the 20 children received assistance from a reading specialist, and 1 child had an Individual Education Plan. However, the multiple strategies we feature can help all students succeed in engineering design.

Up-close view of windmill with four blades.

Engineering Design Process

Engineers use the engineering design process to solve problems (e.g., how to transform wind into useable energy). The nomenclature used to describe the steps of the process varies across engineering education programs. However, the steps take a form similar to those in Figure 19.1 as articulated by Engineering is Elementary (EiE), a national program that has created elementary-level engineering units that link to national science education standards (see Internet Resource). Although Figure 19.1 suggests only one improved design, engineers improve their designs many times. Students can repeat the Improve step as time allows.

Windmill Blade Design

The EiE unit *Catching the Wind: Designing Windmills* was situated within a science unit in which students learned about position, force, motion, and energy. This introductory unit uses inquiry-based science instruction to help students understand these basic physics concepts prior to engagement in the engineering design process. For the culminating lesson in the unit, the students designed windmill blades. The blades are tested on a windmill apparatus placed in front of a fan. Ideally, the wind from the fan causes the blades to turn the windmill's axle, which winds a string on the other end of the axle, lifting a cup (like a bucket-type well). The cup is initially empty; pennies or washers can be added to create a greater lifting challenge. Teachers should operate the fan to ensure that students do not put

FIGURE 19.1.

EiE engineering design process steps and descriptions

Engineering Design Process Step (EiE)	Description of Step
Ask	Identify the problem.
	Determine design constraints (e.g., limitations on materials that can be used).
	Consider relevant prior knowledge (e.g., science concepts).
Imagine	Brainstorm design ideas.
	Draw and label those ideas.
Plan	Pick one idea.
	Draw and label the idea.
	Identify needed materials or conditions.
Create	Carry out the plan; create the design.
	Test the design.
Improve	Reflect on testing results.
	Plan for, create, and test a new (improved) design.

fingers in or get long hair too close to the fan. Students and teachers must wear safety goggles or safety glasses, have long hair tied back to prevent entanglement in the fan blades, and wear only closed-toe shoes or sneakers during windmill operation. OSHA regulations and best practice require fan blades to be protected with a guard having openings no larger than ½ in. (1.27 cm). The teacher should review safety precautions and proper equipment use technique with students prior to starting the activity.

Ask

Students were introduced to the concept of constructing a windmill when the teacher read aloud the EiE storybook *Leif Catches the Wind* (see Internet Resources). In this story, two children, guided by the advice of a mechanical engineer, designed a windmill that turned a paddle to aerate a pond. The teachers noted that the students would design a similar windmill, yet theirs would lift a cup instead of turning a paddle.

To begin the Ask step, the teachers showed the students the windmill apparatus and asked how well-designed blades could make it work. Here and throughout the engineering design process, they encouraged students to use words like *spinning, force,* and *kinetic energy* to describe the windmill's operation and reinforce science learning.

The teachers reminded the students of a sailboat activity the students had completed. In this

activity, children designed simple sails affixed to craft stick masts made from a range of simple materials including felt, paper, plastic bags, and coffee stirrers. The masts attached to sailboat hulls on a low-friction slide and were placed in front of a fan. The sailboat's motion was analyzed for its speed and consistency of motion, then children improved sails based on testing results. The engineering design process was at work here only implicitly; the careful work of moving through each step of the process was saved for the windmill design.

The teachers elicited what the students learned from the sail activity: that certain materials worked better than others, and that large, well-supported sails were effective at catching the wind. They also shared with the students that windmill blades could be designed using the same materials as were used for sail design.

The classroom teacher led a whole-class discussion to complete a worksheet summarizing key aspects of this Ask step: the purpose of the windmill, testing procedures, and what was learned about effective sails. Worksheet questions projected on an interactive whiteboard enabled her to guide struggling writers as she carefully recorded and projected essential Ask-step information for all to document. Students wrote, for example, that the purpose of the windmill was "to catch the wind and lift weight."

Imagine

Students worked in pairs during the Imagine and subsequent steps. During the Imagine step, students brainstormed what different windmill blades might look like that could solve the problem. The focus of this step was on blade shape. An Imagine step worksheet provided four blank boxes in which students draw up to four windmill blade ideas. No scoring rubric was used, but teachers circulated the room to ensure that each pair of students drew at least two ideas and that each idea was drawn with clarity. Students drew a range of blade shapes, including squares, triangles, and rectangles, all affixed to skinny upright rectangles representing craft stick base supports.

Plan

During the Plan step, students selected one of their brainstormed ideas, sketched it, determined how many blades to include in their design, and listed needed materials. The teachers asked students about students' plans and why those plans might help solve the engineering problem. They inquired about students' choices in blade shape and material (Were these choices consistent with what was learned from the sails activity?); the number of blades students chose to put on the windmill (Why 3, 4, or 10?); and, blade angle, whether students chose to angle some or all of

the blades (What made you decide to put this windmill blade at this angle?). The minimal writing demands of this task (the materials list) were primarily supported by the pairing strategy used, in addition to monitoring and assistance from the teachers. When pairing students, a somewhat stronger reader/writer was matched with one who needed a bit more help. This provided peer support for those with reading and writing difficulties. Children who had difficulty listing the materials they planned to use were able to copy from their partner's list after agreeing on necessary materials.

Create

The students eagerly and swiftly created their blades. To prevent shoddy construction, at least eight minutes had to be spent building.

Two pairs of busy engineers seated at one table were Beth and Caitlin (girls) and Trevor and William (boys). One of these children has significant language challenges. Although we refer to these as the "girls" and "boys," we do not intend to generalize their experiences to all girls or boys.

Beth and Caitlin brought their first design to the apparatus. When the wind blew, nothing happened but a quiver of the four large card stock blades set on craft sticks, rattling in place but not spinning the windmill axle. The girls had learned from the sails activity that the blades needed to be large and stiff. However, the blades were not angled. Blade angle was not important in the sails activity, yet it was important for windmill blade design. Students were to discover this essential design feature themselves or with guidance from teachers.

William and Trevor's first design spun the axle and lifted the empty cup. Adding washers to the cup, however, brought the windmill to a halt. Most of their eight windmill blades made of cardstock were angled, yet some faced opposite one another, negating their potentially helpful angled positions. Unable to come to consensus about blade size and shape, the boys compromised: four of the blades were small and triangular (William's idea), and four were somewhat larger, elongated trapezoids (Trevor's idea).

Many of the students at work building at their desks would routinely pause to watch another pair's test. Once all students had tested their first designs, the teachers led a whole-class discussion so that students could share and reflect on their findings with one another.

Improve

The Improve step began immediately after each pair tested their first design, as the teacher asked each pair to reflect on testing results, and continued with more reflection during the class discussion. The students returned to pair work, this time to focus on design improvement. The enrichment teacher circulated, posing questions about how students could improve their designs. She referenced the concept of "form fits function" from an earlier biology unit to suggest that students consider how the features of their designs (e.g., blade shape, size, and material) should be purposeful.

With Caitlin and Beth, who seemed stuck regarding how to proceed, the teacher shared images of real windmills and wind turbines, noting the angle of the blades. She demonstrated how to place one of the blades at an angle by picking up a blade and angling it, and left the girls to modify the other blades.

The enrichment teacher met with the boys and inquired about their first design, yet the boys were satisfied with their blades and the

compromise they had made. The teacher shared the windmill and wind turbine images with the boys, focusing on the blade-to-blade consistency of angle, and used her hands to show the boys how some of their blades opposed one another or were not angled.

After documenting the testing results and improvements they made, both groups were ready to test their improved design. The boys retested and were able to lift 30 washers in the cup, a marked improvement beyond their first design, which lifted only the empty cup. The girls, whose first design did not lift the cup at all, had looks of surprise and joy on their faces as their improved design lifted 50 washers. We suspect that the girls were not only excited by their design success, but felt a rare moment of achievement above others in the classroom.

Assessing the Design Process

During each step of the design process, the teachers formatively assessed students' progress by examining their participation, worksheets, and blade construction and testing. By the end of the design process, all student pairs were successful in lifting at least 20 washers, and many pairs lifted as many as 50.

Summative assessments included the success of students' improved designs and student achievement on a postunit EiE-developed test. It asked questions such as, "Shara is making a windmill, but cannot make it spin. She made the blades bigger, but it still did not spin. Which of the following things could she do to improve her windmill?" (Check all that apply.) Students should correctly identify that Shara could add more blades, change the angle of the blades, or change the materials the blades are made of. Students should leave unchecked "Put holes in the blades to let air through."

Students performed significantly better on the postunit questions that assessed student understanding of wind energy, windmill operation, and engineering than for identical questions on the preunit test. The postunit average was 6.7 of 9 answers correct (standard deviation = 1.3); the preunit average was 4.8 (standard deviation = 1.6).

Helping All Students Succeed

Caitlin, Beth, Trevor, William, and their classmates all met varying degrees of success during the engineering design process. They enthusiastically participated in and documented the steps of the engineering design process. They created a first design using at least some aspects of prior knowledge from the sail activity. All of the pairs were then able to improve their designs, with some having more well-reasoned improvements and ideas than others. Online, we have listed teaching strategies introduced in the vignette that encourage success for all students in engineering design, especially those who have special needs (see NSTA Connections).

Although we want well-reasoned designs, we also want students to try their own ideas. If students are attached to design ideas that fail to incorporate relevant concepts, allow them to test their designs. Testing results will help make the case to students that the most successful designs employ good reasoning.

We end with two warnings: (1) brace yourself for the excitement that students have as they engage in the engineering design process; and (2) be prepared for all students to succeed and for some who normally struggle to shine.

Acknowledgment

The Workforce One Maryland Program through the Maryland State Department of Labor, Licensing, and Regulation, generously funded this project.

NSTA Connection

Download a list of teaching strategies at *www.nsta.org/SC1003*.

Internet Resource

Engineering Is Elementary (EiE)
   www.mos.org/eie

CHAPTER 20

Fuel-Cell Drivers Wanted

By Todd Clark and Rick Jones ///

I f you were to ask middle level students what type of car they would like to drive when they get their driver's license, they would no doubt respond with "a sports car" or an SUV. Although they are years away from being licensed drivers, most have spent considerable time contemplating their dream car. Suggesting that maybe they should consider a fuel-cell car or a hybrid vehicle may generate some laughs, but may also pique their curiosity.

Former President Bush once commented that the greatest environmental progress made in this century will be through technology and innovation (Bush 2003). His administration committed $1.7 billion for research on the use of fuel cells for personal transportation (Abraham 2003). While the political climate is favorable for the development of fuel-cell vehicles for personal transportation, the market's demand may not be so favorable. Nonetheless, middle level students will be the next generation of drivers and voters, and they need to be able to make informed decisions regarding the nation's energy and transportation policies and understand how such decisions impact our environment.

Fuel-Cell Technology

Fuel cells generate electricity by the reverse reaction of electrolysis of water. In the electrolysis, electrical current causes water molecules to

FIGURE 20.1.

Chemical equation

The combined chemical reaction for fuel cells is simply:

$$2H_2 + O_2 \rightarrow 2H_2O \qquad (1)$$

This is the combination of two other chemical reactions, one at the cathode:

$$O_2 + 4H^+ + 4e^- \rightarrow 2H_2O \qquad (2)$$

and one at the anode:

$$2H_2 \rightarrow 4H^+ + 4e^- \qquad (3)$$

FIGURE 20.2.

Critical-thinking questions

- What would be the advantages and disadvantages of hydrogen as an "energy currency" relative to our current use of carbon-based fuels for energy storage?
- There was some expectation that battery-storage electric vehicles would be the next generation of personal transportation. Why does that seem unlikely today? How are fuel-cell vehicles different from electric vehicles and how are they similar? Do the same drawbacks that apply to electric cars with batteries apply to fuel-cell vehicles?
- Fuel-cell cars will initially be more expensive than gasoline-powered cars. How should consumers decide if the additional expense is warranted? Should the government provide incentives for drivers to purchase alternatively fueled vehicles? Why or why not?
- The infrastructure of gasoline stations throughout the country to provide fuel for gasoline-powered cars was built up over many decades. How would an infrastructure to provide hydrogen to fuel-cell cars be different and is there a best method for making this transition, if it occurs?

separate into their component gases: two parts hydrogen and one part oxygen. In a fuel cell, hydrogen combines with oxygen to produce electricity, heat, and water (see Figure 20.1 for the chemical equations used in a fuel cell).

The U.S. Department of Energy (DOE) has committed resources to help teachers and the general public understand the science behind this new technology and the benefits and disadvantages of fuel-cell technology (see critical-thinking questions in Figure 20.2).

Start Your Engines

As an initial introduction, it may be helpful to have the students brainstorm what they know about hydrogen, the sources of hydrogen on Earth, and its potential uses. Ask students if it is possible to race a model car using only distilled water as fuel or how NASA provides safe electrical power aboard the space shuttle.

The idea for using hydrogen and oxygen to produce electricity has been around for about 100 years and practical use of fuel cells to provide electrical power for spacecraft has been a reality for more than 40 years. Students may be provided with the materials necessary to design and carry out several experiments that will allow them to separate hydrogen and oxygen from water. From these explorations, students will have the necessary conceptual understanding of electrolysis.

More Than Just a Car Race

In May of 2003, in conjunction with the DOE's National Science Bowl, 50 students from around the country participated in the First Annual Fuel-Cell Car Challenge at the National 4-H Center in Chevy Chase, Maryland. Students had approximately 12 hours to design, construct, and race their cars in two events: driving up an inclined plane with increasingly steeper angles and a head-to-head, double-elimination speed race. In this competition, students were provided key components, such as a fuel cell, in addition to other materials like balsa wood and juice boxes (for adapting this competition for use in your classroom, see the complete list of materials in Figure 20.3).

Students designed vehicles that were as varied as their personalities, and the time limit was still long enough to allow for multiple designs and redesigns as students explored their understanding of gear ratios, torque, friction, and Newton's laws of motion. The winning team for the inclined plane test used impressive gear ratios (1,300:1) to climb a 44-degree incline, while the winning team for the speed race used a three-wheel design (to

FIGURE 20.3.

Materials list

- Thames and Kosmos Hydrogen Fuel-Cell Car Kit (plastic car body with water reservoir removed)
- One empty juice box (with 10 cm × 3 mm plastic straw)
- One 6-speed gearbox kit (Tamiya)
- One bag "OO" gears, shafts, and accessories for 2 mm shaft (Solar World)
- Two high-speed, 2.54 cm diameter racer wheels for 3 mm axle with matching O-ring (Pitsco)
- Two 5 cm diameter pulley wheels for 3 mm axle with matching O-ring (Sargent Welch)
- Four brass sleeves (3 mm × 1 cm)
- One 30 cm × 3 mm brass axle rod
- Four alligator clips
- Two 30 cm length pieces of 20-gauge copper wire
- One bottle instant bond cement
- Four 5 mm cable staples and 4 small wood screws
- One roll electrical tape
- One utility knife
- One set small (jeweler's) screwdrivers
- One 15 cm × 50 cm × 0.6 cm piece foam core
- One 15 cm × 30-cm × 0.6 cm piece balsa wood
- One 7.5 cm × 46 cm × 0.6 cm piece balsa wood

reduce friction) with a long, narrow chassis to help the car align quickly with the guide wire.

The competition, while exciting, was secondary to learning about how fuel cells operate, engineering design principles, and cooperative problem-solving skills.

Driving Is Learning

By building and racing these cars, students learn firsthand two advantages of using fuel-cell technology: (1) the fuel is hydrogen, which is the most abundant element in the universe and (2) the product of the internal reaction is water (see Figure 20.2), which is far more benign than the many products of the incomplete combustion of hydrocarbons.

Students also discover that another advantage of using a fuel cell to generate electricity is that there are no moving parts, making it more efficient than a generator. Like batteries, fuel cells can be connected in series and in parallel to meet different electrical loading requirements. In addition, the output of a fuel cell—water—can be converted back into hydrogen and reused as fuel.

Competitions with fuel-cell-powered model cars can help teachers integrate several science areas. Students learn about chemistry when they study the chemical reactions that produce electricity and water in fuel-cell operations. Students apply their knowledge of physics when they study gear ratios, friction, Newton's laws, and the center of gravity in constructing their model cars. Students engage in engineering when they work as a team to design, modify, redesign, and test their vehicles. Finally, students learn about communication when they work with teammates to develop a division of labor that will allow them to construct their cars in a timely manner.

References

Abraham, S. 2003. Remarks prepared for Secretary of Energy Spencer Abraham at International Energy Agency Ministerial Working Dinner, Monday, April 28, 2003, Le Meridian Etoile Hotel Paris, available at *www.evworld.com/databases/shownews. cfm?pageid=news290403-06*

Bush, G.W. 2003. Comments made by President George W. Bush at the National Building Museum, February 6, 2003 *www.white house.gov/news/ releases/2003/02/20030206-12.html*.

Internet Resources

DOE fuel cell information
    www.eere.energy.gov/hydrogenandfuelcells/ education.html
How fuel cells work
    http://science.howstuffworks.com/fuel-cell.htm
Junior Solar Sprint Rules
    www.nrel.gov/education/student/natjssrules.html
Model fuel-cell car kits
    www.thamesandkosmos.com/store/fuelcell1.html and *www.heliocentris.com/products/school.html*
National Middle School Science Bowl
    www.scied.science.doe.gov/nmsb/default.htm

CHAPTER 21

The Science of Star Wars

Integrating Technology and the
Benchmarks for Science Literacy

By Stephanie Thompson //

*S*tar Wars: The name alone implies action, adventure, the vastness of space, alien creatures, and of course, who can forget lightsabers? The *Star Wars* saga has entertained millions of people around the world for years, so I asked myself: Why couldn't *Star Wars* motivate a group of students to delve into the depths of technology?

I decided to assign my students the task of designing and building a working lightsaber. A lightsaber is a fictional weapon created for the *Star Wars* movies. The lightsaber consists of a handgrip similar to a flashlight and a tube of colored light that forms the blade.

Mentioning the mythical weapon was all it took to send a group of kids off into a frenzy of collaboration. I had never seen this class show more motivation and excitement over a project. Who thought school could be fun?

The students were instructed to design, build, and explain the scientific reasoning behind their projects based on standards taken from the *Benchmarks for Science Literacy* (AAAS 1994). The project aligns with the curriculum in

that it allows students to demonstrate the appropriate use of the science process skills and use of technology in the classroom.

Getting Started

In the first few days of the project, the students were working and collaborating with one another to discuss design ideas for the lightsaber. Students had great ideas for assembling their own lightsaber, but found it difficult to create a set of detailed written instructions based on their ideas. I spent some time in class discussing methods of writing appropriate instructions, and provided appliance user manuals downloaded from the internet as examples of how instructions should be written.

I placed students into groups and had them compare the sets of instructions from the appliances. Students reviewed several sets of instructions and wrote down the most helpful information in each. Students found that the instructions were easier to follow if they were broken into small steps and included pictures or diagrams. We finished with a class discussion about what makes a good set of instructions. Students then practiced writing instructions for constructing their lightsabers. Once the instructions were written, they shared with several classmates to obtain feedback. (Sample student instructions are available with the online version of this article, available at *www.nsta.org/middleschool*.)

When the instructions for building lightsabers were finalized, we began construction. Students were asked to bring in their own supplies from home, and the resulting flood of materials turned my classroom into a hardware store. I am lucky to have a classroom and separate lab, so students were able to keep their supplies in grocery sacks in the lab. I supplied basic tools, glue, tape, scis-

BUILD YOUR OWN LIGHSABER

Long, long ago, in a galaxy close to home, there was a teacher who challenged her students to build lightsabers. The apprentice builders were known as *padawan* and their teacher was known as Thompson-Wan Kenobi.

Students were instructed to design and build a model lightsaber, 30 cm in length. The lightsaber blade could be any color, but the blade had to light up. In order to complete their task, students had to use the "Scientific Force" to guide their work, while fighting off the urge to turn to the "Dark Side" and not follow Thompson-Wan Kenobi's instructions. A sad fate awaited those students who turned to the Dark Side.

In order to become a Jedi, each padawan was required to:

- Complete a set of designs for the lightsaber.

- Write a set of instructions for the lightsaber.

- Write a persuasive paper detailing the pros and cons of developing a lightsaber and its possible impact on society.

Project Timeline

Day 1: Student groups formed and project outlines distributed.

Day 5: Completed designs and instructions due.

Day 8: Materials needed to build lightsaber brought to class.

Day 15: Completed lightsaber due.

Day 18: Persuasive paper due.

Class Discussion Questions

1. Is it possible to develop an actual lightsaber or is it truly only science fiction?

2. Should this weapon be created? What kinds of effects would a weapon such as the lightsaber have on society?

3. Would this be a suitable weapon for the military? What would happen if this kind of technology got into the wrong hands? Should this technology be shared with the world?

4. Would the creation of the lightsaber change the quality of life for some people in a negative or positive manner?

sors, measuring tapes, batteries, wiring, and mini lightbulbs. Once the materials were collected, students quickly began assembling their lightsabers.

Management

To minimize the mess and allow students to assist one another, I divided the class into groups of four. I allowed students a week of class time to construct their lightsabers, but most students were finished within three, 45-minute class periods. For multiple classes, it would be helpful to separate communal supplies into separate storage tubs in the lab. Sort the communal supplies in large freezer bags for each group. If the group runs out of supplies, then students could ask another group for a particular item or trade with another group. My students were very cooperative and problems with communal supplies never arose.

If students were unable to bring supplies from home, I supplied them with the needed materials. I was able to obtain many donated materials from local stores and businesses, such as electrical contractors. Parents also responded to a call for supplies that I posted on our school website. On average, the cost for the materials needed to build a lightsaber was around $10.

Finding the appropriate light source for their lightsabers was the biggest challenge for my students. Some of the kids used flashlights for their light source; others used a more complicated design, building their own circuitry system. After students decided what light source to use, they selected tubing that would form the "blade" portion of the lightsaber. Students used various materials to create the blade, including clear plas-

tic tubing, PVC pipe, and old toy parts. The lightsabers came to life in a variety of colors achieved using colored cellophane, plastic wrap, and tape. The completed lightsabers featured a dazzling array of designs, colors, and sound effects. We finished the project with some class discussion questions designed make students think about the science, technology, and society issues related to lightsabers.

The Persuasive Paper

After the lightsabers were completed, students were assigned the task of writing a two-page persuasive paper in which they explained the value of this project. I provided students with quotes from the *Benchmarks for Science Literacy* for inspiration. Figure 21.1 (p. 152) provides the specific benchmarks as well as samples of how students might use them to justify this project. I graded the papers using the rubric in Figure 21.2 (p. 154).

Conclusion

This project has helped me incorporate technology into the classroom, strengthen students' procedure writing skills, and raise issues of how science affects the world around us. The *Benchmarks for Science Literacy* can be a valuable tool for improving classroom practice, and this project is a great way to engage students and generate enthusiasm for science.

Reference

American Association for the Advancement of Science (AAAS). 1994. *Benchmarks for science literacy.* New York: Oxford University Press.

FIGURE 21.1.

Benchmarks for persuasive paper

The Scientific World View

1. Scientists differ greatly in what phenomena they study and how they go about their work. Although there is no fixed set of steps that all scientists follow, scientific investigations usually involve the collection of relevant evidence, the use of logical reasoning, and the application of imagination in devising hypotheses and explanations to make sense of the collected evidence.

   How it relates: Students must perform a scientific investigation and apply their imagination to build their lightsaber.

2. Computers have become invaluable in science because they speed up and extend people's ability to collect, store, compile, and analyze data; prepare research reports; and share data and ideas with investigators all over the world.

   How it relates: Students must use computers in this project "to collect, store, compile, and analyze data, prepare research reports, and share data and ideas.

The Nature of Technology

1. Technology is essential to science for such purposes as access to outer space and other remote locations, sample collection and treatment, measurement, data collection and storage, computation, and communication of information.

   How it relates: Students must use computers in this project as a form of communication.

2. The human ability to shape the future comes from a capacity for generating knowledge and developing new technologies—and for communicating ideas to others.

   How it relates: Students must think about the future of technology. Is it possible to develop an actual lightsaber or is it truly only science fiction? Should this weapon be created? What kinds of effects would building a weapon such as the lightsaber have on society?

3. Technology cannot always provide successful solutions for problems or fulfill every human need.

   How it relates: Students must think about the possibilities of building a lightsaber. Would this be a suitable weapon for the military? What would happen if this kind of technology got into the wrong hands?

4. Technology has strongly influenced the course of history and continues to do so. It is largely responsible for the great revolutions in agriculture, manufacturing, sanitation and medicine, warfare, transportation, information processing, and communications that have radically changed

how people live. New technologies increase some risks and decrease others. Some of the same technologies that have improved the length and quality of life for many people have also brought new risks.

How it relates: Again, students must think about the consequences of the lightsaber in the usage of warfare and how it might affect the way people live. Would the creation of the lightsaber change the quality of life for some people in a negative or positive manner?

5. Rarely are technology issues simple and one-sided. Relevant facts alone, even when known and available, usually do not settle matters entirely in favor of one side or another. That is because the contending groups may have different values and priorities. They may stand to gain or lose in different degrees, or may make very different predictions about what the future consequences of the proposed action will be.

How it relates: Students must think about how people will react to the construction of a real lightsaber. Will politics come into play? How might different groups feel about the creation of the lightsaber?

6. Societies influence what aspects of technology are developed and how these are used. People control technology (as well as science) and are responsible for its effects.

How it relates: Students must think about the development of the lightsaber and how societies must be held responsible for deciding whether or not to create a lightsaber.

The Designed World

1. The choice of materials for a job depends on their properties and on how they interact with other materials. Similarly, the usefulness of some manufactured parts of an object depends on how well they fit together with the other parts. Manufacturing usually involves a series of steps, such as designing a product, obtaining and preparing raw materials, processing the materials mechanically or chemically, and assembling, testing, inspecting, and packaging. The sequence of these steps is also often important.

How it relates: Students must think about the materials needed to build a lightsaber and how those materials will work and fit together.

FIGURE 21.2.

Star Wars persuasive paper rubric

Points	Appearance
0	Illegible, not typed, ripped or torn paper, no cover page, no references
1	Paper is typed in 12-point font, but ripped, no cover page, no references
2	Paper is typed in 12-point font, but there are many misspellings, cover page and references present but incomplete
3	Paper is typed in 12-point font, fewer than three misspellings, complete cover page, references present
4	Paper is typed in 12-point font, no misspellings, presentation is professional, complete cover page, references are listed
	Research
0	Incomplete research, no detail, lacks clarity, incomplete sentences
1	Report does not include all information from research, complete sentences, no details
2	Research is given with some detail, all parts of the research are included
3	A thorough and complete report is given, capturing the attention of the reader, all of the research is included
4	Research is complete, capturing the attention of the reader, extra details are included with examples
	Grammar
0	No punctuation, commas, capital and lowercase letters, indention of new paragraphs, and so on
1	Lacks some punctuation, commas, capital and lowercase letters, indention of new paragraphs, and so on
2	Punctuation, commas, capital and lowercase letters, indention of new paragraphs, and so on are present
3	Punctuation, commas, capital and lowercase letters, indentions of new paragraphs, and so on, are present and used appropriately
4	Punctuation, commas, capital and lowercase letters, indentions of new paragraphs, and so on are used appropriately and effectively. Excellent grammar usage
	Scientific Analysis
0	No scientific analysis applied to research
1	Little scientific analysis applied to research
2	Some scientific analysis applied to research
3	Thorough scientific analysis applied to research
4	Excellent scientific analysis applied to research

Point Value:　　0–5 = 60　　　6–8 = 70　　　9–11 = 80　　　12–14 = 90　　　15–16 = 100

CHAPTER 22

Catapulting Into Technological Design

By Kristen Hammes //

One of my favorite projects for exploring levers is to have students design and build a working catapult. The catapult is designed at home and built from scratch during class. Each catapult is tested for accuracy and students determine the average speed of the projectile (a marshmallow), as well as the amount of force, work, and power involved in the launch. On the final day of the project, students use their catapults to knock down a castle made from empty soda cans.

The project begins with a short introduction to first-, second-, and third-class levers. I follow up by identifying the fulcrum, effort, and load found on common levers in the room, such as a pair of scissors. Scissors are a first-class lever with the effort applied to the handles, the fulcrum in the middle at the screw, and the load placed in between the blades.

Next, I ask students if they have ever seen a catapult. Students will then recall examples they have come across in video games, movies, and social studies class. If possible, I try to schedule this project at the beginning of the year when

the Roman Empire—a catapult innovator—is being discussed in social studies. I then point out that many catapults are levers and identify the fulcrum, effort, and load on a diagram of a typical model.

Once everyone is familiar with catapults and the principles behind their operation, I hand out the activity sheet and explain that students will be building their own siege engines. Students can work on their own or in small groups for this project. Before construction begins, they need to create a scale drawing of their device with measurements noted. To get them started, I include a number of websites on the activity sheet that provide plans for catapults. I tell students that these

CONSTRUCTING A CATAPULT

You will be constructing a catapult in class using materials provided and found in your home. The catapult must meet the following design criteria:

- Catapults must be examples of a lever: no slingshots are allowed.

- Catapults must be able to fit through the door to the classroom, and you must be able to carry your catapult yourself.

- Catapults must shoot a small marshmallow a distance of 5 m. If it shoots too far or not far enough, you must be able to make adjustments.

- No parts may come off your catapult when it is launching. If something does fly off, it may not be replaced.

- All blueprints must be drawn to scale and actual measurements must be included on your sketch.

- You must build the catapult according to your blueprint; any adjustments must be noted on the blueprint.

- The fulcrum, effort, and load of the catapult must be labeled on the blueprint.

Sample blueprint plans can be found at *icatapults. freeservers.com/plans.htm* or *www.knightsforhire. com*. These are to serve as models and should not be copied directly.

Safety

- Students must wear goggles at all times during the construction and testing of the catapults.

- Keep your work area clear of backpacks, books, and other clutter during construction.

- Keep tools and materials on tables and off floors to avoid creating tripping hazards.

Postconstruction Questions

1. How can you adjust your catapult to hit a target that is not a constant distance from you?

2. What will happen to the firing distance when the length of the load arm changes?

3. What will happen to the firing distance when the load arm is pulled back more or less than on previous shots?

Calculations

1. Use a spring scale to measure the force required to launch the marshmallow. For most catapults, this requires hooking the spring scale to the load arm (in between the fulcrum and the load), and pulling back the arm to the load position.

2. Measure and record the distance the load was pulled back in centimeters and convert to meters for easier calculations.

3. Multiply the force by the distance to determine the work done to launch the marshmallow.

4. Use a stopwatch (or other electronic measuring system) to measure the time it takes to launch the marshmallow—from load position to release of projectile.

5. Divide the work by time to determine the power of the launch.

6. Measure the distance the projectile travels by the time it is in motion to determine the average speed of the projectile.

Force used to launch = _____ newtons Distance load pulled back = _____ meters

Work (f x d) = _____ joules
Time to launch = _____ seconds

Power (W/t) = _____ watts
Average speed (d/t) = _____ meters/sec

plans should only serve as inspiration, however, and should not be copied. Students generally work on their designs for homework over two evenings.

There are a number of design requirements that are included on the activity sheet, but the most important are that the catapult is a lever and must be adjustable. The ultimate goal for stu-

Rubric

Score	Blueprint	Catapult	Accuracy	Thinking it through
5	• Drawn to scale • Measurements are included and are accurate • Sketch is neatly drawn and accurately represents finished catapult • List of materials is complete and has been initialed by parent • Fulcrum, effort, load and class of lever are identified	• Catapult is completed. • Catapult is a lever. • Fulcrum, effort and load are correctly pointed out. • Class of lever is correctly identified.	• Projectile hits 90–100% of targets positioned at varying distances within three attempts.	• All postconstruction questions have been answered. • Student can discuss findings and can show how adjustments are made to the catapult.
4	ONE of the above items is missing.	• Catapult is completed. • Catapult is a lever. • Fulcrum, effort and load are correctly pointed out (may take a second attempt). • Class of lever is correctly identified.	• Projectile hits 80–89% of targets positioned at varying distances within three attempts.	• All postconstruction questions have been answered. • Student can discuss findings and can show how adjustments are made to the catapult (may reference written work).

Integrating Engineering + Science in Your Classroom

Rubric (continued)

Score	Blueprint	Catapult	Accuracy	Thinking it through
3	TWO items are missing.	• Catapult is completed (finishing touches added within this class period). • Catapult is a lever. • Fulcrum, effort and load are correctly pointed out (may take a second attempt). • Class of lever is correctly identified (may take a second attempt).	• Projectile hits 70–79% of targets positioned at varying distances within three attempts.	• Most postconstruction questions have been answered. • Student can discuss findings and can show how adjustments are made to the catapult (may reference written work). • Student needs to think through answers before responding.
2	THREE items are missing.	• Catapult is nearly completed (can be completed within the class period). • Catapult is a lever. • Fulcrum, effort, and load are correctly pointed out (may take a second attempt). • Class of lever is correctly identified (may take a second attempt).	• Projectile hits 60–69% of targets positioned at varying distances within three attempts.	• Some postconstruction questions have been answered. • Student has difficulty explaining answers and how to make adjustments to catapult. • Student needs to ask others for help in order to respond.
1	FOUR items are missing.	• Catapult is not completed. • Catapult is not a lever. • Student is unable to identify the fulcrum, effort and load even with teacher help. • Student cannot identify the class of lever.	• Projectile hits less than 60% of targets positioned at varying distances within three attempts.	• Postconstruction questions have not been answered. • Student cannot explain answers and how to make adjustments to catapult. • Student needs to ask others for help in order to respond.

dents is to launch a projectile and strike a target 5 m away. Whichever catapult design students choose, they must be able to make adjustments to the device to control the distance the projectile travels.

Once the designs are complete, catapults are brought to class and reviewed by classmates to see if the design requirements have been met and the drawing is complete. Each design should be reviewed by at least two other students or groups. Any problems with the designs should be noted and passed along to the designer. If no problems are found, the design can be initialed and returned.

After the plans have been reviewed, students can make any necessary design modifications and compile a list of materials needed for construction. The teacher reviews and approves all final designs and materials. If possible, materials should be brought from home, and parents should sign off on the list. If a student wants to use power tools, they must be used at home under adult supervision. You could also make arrangements with a shop teacher or handy parent to help with the construction. Encourage students to bring in extra materials and tools to share with others who may not have access to or the ability to buy everything they need. Often, home improvement centers or lumberyards will donate scrap materials.

I set aside two class periods for construction and testing of the catapults. Before construction begins, we review all safety rules. Most importantly, safety goggles must be worn at all times during construction and testing of catapults. Additional safety rules are included on the activity sheet. If possible, have a couple of handy parents volunteer to help out during construction and testing. As students begin, circulate around the room to supervise construction and trouble-

shoot problems. Students who finish early can begin testing in order to make adjustments.

Once construction is completed, students return to the activity sheet to explain how they plan to adjust their catapult and predict how specific variables will affect its operation. Students are now ready to head to the firing range. I set up two firing ranges back-to-back in the hallway, so students are firing their projectiles in opposite directions. Again, I recommend that you have someone on hand to help supervise. At one end of a range I put a piece of masking tape across the hallway floor to create a starting line. Ten meters away I place another piece of tape to mark the end of the range. Between the two, at 5 m, I tape an X on the floor to serve as a target. The starting line for the other range is positioned about 3 m away from the first and the rest of the range is laid out down the opposite end of the hall.

Each student is given one marshmallow to take out onto the firing range, and it must be returned to the teacher at the end of the testing. The marshmallow may not be eaten! Students test their catapult for five minutes and return to class to make any necessary adjustments. When students are ready to return to the range, they add their names to a list on the board and I move testers on and off the range every five minutes.

After students have fine-tuned their catapults, they can begin gathering data using the formulas found on the activity sheet. Measuring the time that it takes to launch the catapult can be difficult without a photogate and electronic timer, but you can attempt it with a stopwatch. If timing the launch proves to be too inexact, you can estimate the related calculations.

We wrap up this project with final presentations: The designers bring their catapults to the front of the room; identify the fulcrum, effort, and load; and explain what class of lever is involved. Students then explain how the catapults can be adjusted and demonstrate by making three attempts to launch a marshmallow into a large plastic tub placed 5 m away. At the end of each presentation, I collect students' activity sheets for grading (see rubric, p. 157).

After individual testing is done, I divide the class into two groups for our marshmallow battle. I set up two bunkers (tables lined up) on opposite sides of the room, where students can set up their catapults. In the center of the room (about 5 m from each front), I build a castle made from empty soda cans. The two teams face opposite sides of the castle. When I give the signal, the two teams (with safety goggles in place) begin their assault on the castle, which lasts until one team knocks down their side of the castle. As the battle rages, I sit back and watch as students adjust their catapults and have a great time demonstrating a hands-on application of technological design.

CHAPTER 23

Potato Problem Solving

A 5E Activity Addresses Misconceptions About Thermal Insulation

By Sarah J. Carrier and Annie Thomas //

"Watch out, the stove will burn you," "Ooh, ice cream headache!" Students construct their conceptions about heat and temperature through their own intuitions about daily life experiences. For example, students often believe that different objects maintain their own temperature, metals being hot and plastics being cool. Misconceptions can be born from these constructed concepts. Paik, Cho, and Go (2007) found that students see quantitative measures of temperature as summative. When asked what would happen when you combine two waters of 30°C each, students replied that the result would be 60°C water.

Our third- and fourth-grade students have expressed the view that one thick blanket would provide more warmth than multiple thin blankets, failing to recognize the contribution of air as an insulator when held within layers of fabric. Thermal properties and heat transmission are quite complex; however, elementary school students can begin to construct basic but sophisticated understandings of heat and insulation through hands-on activities and experiences. Learning about heat and temperature concepts such as conduction, convection, and

insulation serves as a foundation for chemistry, physics, and biology (Paik, Cho, and Go 2007).

The activity described here challenges students to design a structure with good insulation properties. The project provides opportunities to problem solve through exploration, investigation, design, structural analysis, synthesis, and revision. The 5E model (Bybee 1997) of the learning cycle lends itself to this unique problem-solving model. The lesson described here has been conducted with third- and fourth-grade students; however, various adaptations can be implemented for any ability level. The teacher can conduct a whole-class demonstration for younger students' first experiences or provide more autonomy for students of advanced ability levels.

Classroom Context

The activity described here fits well into a unit on the properties of matter. It also gives students experience with experimental design and the connections of science with mathematics, engineering, and technology.

Our school year begins with a measurement unit. Students learned about the difference between weight and mass; this was reinforced when they compared their weights on different planets early in the school year. Before beginning this activity, students should also have been introduced to the importance of controlling variables during investigations.

The first step in this investigation is research, and the level of student autonomy varies depend-

Students test materials to find the best insulator.

ing on the grade level. Students in grades three and four can find key ideas from nonfiction material at their level, with various levels of scaffolding. The teacher can provide age-appropriate reading or internet links to resources that will give students an introduction to heat and insulation (see Resource). Many students are intrigued by penguins, so they make a good topic of discussion. Unlike other birds, penguin feathers are lined up with multiple layers and can alternate position, self-organizing whether the penguin is in water or air. This ability to trap air, combined with a layer of fat under their thick skin, allows penguins to survive in Antarctic conditions (Attenborough 1979). Bioengineers are studying penguins to design effective insulation materials.

Present a Problem

Students are introduced to this lesson by way of a fictitious scenario that engages them and gives them a real-world problem to solve:

Imagine you receive a frantic call from your potato farmer neighbor, Pat. Pat has called you because she knows what an amazing scientist you are and hopes you will find a way to help her out. The good news is that Pat produced far more potatoes than ever before this season. The bad news is that Pat has nowhere to store the potatoes inside, and tonight there is going to be a freeze. Pat does have some materials she could use to build storage boxes, but she is unsure which would be the most insulating. This is where you come in. What type of container will you create? Pat has seven materials to choose from: foam board, aluminum foil, corrugated cardboard, Styrofoam, bubble wrap, felt, and feathers. (Use purchased, sterile feathers; check with the school nurse about student

allergies.). Pat needs these materials for other projects, so she asks that you only use two of the materials to construct the boxes.

New foam board should be aired out for several days prior to using, as it contains volatile organic compounds which can cause allergies. Science fair display boards or pegboards provide allergy-free alternatives. Students must wear safety goggles during the construction.

Many of the materials used in this lesson are familiar to the students and are easily available and affordable for teachers. However, students may not have considered these materials as having insulation properties. It is ideal to have the materials available for students to examine, including measurement tools students may need to solve this problem (stopwatches, thermometers [nonmercury type], and rulers).

The students work in groups of three to develop their strategies and procedures. Teacher-initiated dialogue focused on insulation and its properties will aid students in determining how to begin. Questions such as, "What types of insulation have you seen before?" can probe students to think about building materials such as insulation in an attic or materials used to insulate people, such as blankets, quilts, and jackets. This conversation directs students to look at insulation as a material that limits or stops the transfer of heat.

After introducing the property of insulation, have students consider the materials. "How can you determine which of these materials will be most efficient at insulating your potato?" Student ideas are investigated in the explore phase.

Research Through Investigation

During the explore phase, students are challenged to determine which materials would be most suitable for their potato insulation boxes. To do this, students must devise an investigation to test the properties of the materials and determine how they will quantify their results. Asking good, productive questions, such as "How could we measure which material will keep out the cold the best?" will steer students toward collecting quantitative data. There are numerous ways to test these materials effectively, but most often students agree upon the following:

1. Read and record the room temperature in Celsius on the thermometer.

2. Place a frozen ice pack on the work surface.

3. Place a single layer of one of the materials on top of the ice pack. Use caution in handling frozen ice packs, as they can cause "cold" burn on skin. Do not handle packs directly.

4. Place a thermometer on top of the material.

5. Use a stopwatch to time five minutes.

6. After five minutes, record the temperature.

7. Wait until the thermometer has returned to the prerecorded room temperature.

8. Repeat steps 2–7 with each material.

This is one of many methods that may be used for testing the insulation levels of the materials. Depending on how many resources are available in any given classroom (e.g., number of thermometers, insulation materials) the teacher may choose to allow each group to come up with its own testing method or the class could agree on a single method and test as a whole. There are benefits to both methods. Allowing each group to determine its own method of testing the materials may reveal novel testing methods that the teacher never considered. When testing as an entire group, each suggested method can be presented and discussed, which would offer an

opportunity to spark a dialogue on the need to control variables, collect quantitative data, and the use of other science-process skills.

Student groups record everything in their science notebooks. Depending on their ability levels, the teacher can provide varying levels of scaffolding, such as premade data collection charts. As students build experiences with organizing data, they develop their autonomy as they learn to design their own charts.

Student groups now use their research to make informed decisions about which two materials they will choose to construct their potato insulation boxes. Students describe which of the two materials will form the base layer and how to measure equal amounts of the different materials. Students are given the parameters of length, width, and height for the box that holds one potato (20 cm × 10 cm × 10 cm) to reinforce consistency of variables. Students should find the mass of the potatoes to ensure that a similar size potato is placed in each box.

Some of the materials may be difficult or dangerous for a child to cut. If so, have students measure and mark lines—and cut the materials for them. A good option is to offer duct tape as a means to hold the pieces of each box together. It is important to give each student group only 1 m (100 cm) of tape, as some groups may use too much and thus add another layer of insulation to their potato box.

After the boxes are constructed, it is time to prepare the potatoes. A teacher should assist by using a knife to score the potato and then sticking the thermometer into the center. The depth of the thermometer is measured to assure consistency, another variable held constant. A large cooler with ice is necessary, as it is important to place all boxes in a single row. Once the potatoes with thermometers are all inside the boxes, the cooler is closed. After two hours, students record the temperatures of the potatoes in each box.

Design and Construct a Solution

We start the explain phase by having students share their findings with the class. A large chart on the blackboard or chart paper that is visible to all students may be used to compile the data from each box. Students collect data from other groups and develop a chart and a graph to communicate their data and share it with others.

It is important that the students maintain a record in their science notebooks, including detailed documentation of their measurements and design. The emphasis on detailed record keeping supports the importance of precise communication so other scientists can use the descriptions to replicate the investigation. Their detailed documentation also helps them remember from trial to revision.

Students record their designs using digital photography to supplement their drawings and descriptions in their notebooks. Students take turns with digital cameras from our school's media center and take photos during the construction process and of the finished product. The photo images allow them to compare and contrast investigation strategies and their subsequent revisions, providing them with a concrete record of their efforts and procedures.

Following students' attempts to communicate their findings, teachers help students make sense of their observations. Building on the activity and student discussions, the teacher formally introduces the scientific concepts. Key words are displayed visually to supplement the oral interactions. We remind students that heat naturally flows from warm to cooler things, and we relate the topics to their real-world experiences. Their experiences could include clothing used to stay warm in winter, using sleeping bags at camp, or metal spoon handles becoming warm when placed in a cup of hot chocolate. We review the concept that anything blocking the flow of heat provides insulation. These key concepts are related back to the students' box design activity as they discuss patterns in insulation choices and heat retention among the groups.

Revise and Retest

Investigation should not end when students have learned new information. In the elaborate phase, students have the opportunity to revise their design solutions and retest, providing another learning opportunity and a chance to assess student learning.

Student groups are asked to examine the data they collected from both the initial materials testing and their design tests. They can then choose to rebuild their potato insulation boxes using other materials or retest their original boxes.

The following activity provides students additional opportunities to make qualitative descriptions of their personal sensations as they compare insulation materials. Two zippered sandwich bags are zipped together with various insulation materials in between the layers of the two bags. This creates an open space that students can slip their hands into, surrounded by insulation. A second pair of bags is zipped to each other without insulation between the layers. The students first make predictions about which materials will provide more insulation, then they submerge both hands into tubs of ice water. They compare the sensation of cold of their hand in the bare zippered bag with the sensation of the insulated zippered bag on the other.

Sharing Results

Students' detailed records of their initial designs and insulation materials are helpful in this evaluate phase. Students' graphs and charts as well as descriptions of conclusions are used to assess their understandings about insulation, data collection and recording, measuring, and experimental design.

This is a perfect time for the scientific process skill of communication to be modeled. Student groups prepare presentations to communicate their designs and the results using their digital photos and presentation software. When students report their trials to the other scientist designers in the classroom, sharing their research designs and procedures as well as what worked as expected and what surprised them, they participate in scientific discourse. The project provides a novel problem-solving experience in science and an authentic form of performance assessment. This allows us to test not only their knowledge of the properties of insulation, but also the students' developing understandings of experimentation processes. The performance assessment that measures their procedures, revisions, and analysis of findings supplements the assessment of their science notebook entries for recording data and drawing conclusions.

Potato to STEM

This single lesson incorporates many of the goals of STEM instruction by encompassing learning experiences in science, technology, engineering, and mathematics. In addition to using presentation software to share their investigations, students incorporate technology with digital cameras and computer-generated charts, graphs, and text.

Many adaptations are possible with this activity, including variations in the amount of scaffolding based on the students' grade and ability levels and providing a variety of construction materials. For example, the teacher could provide a template for construction of the structure. This would allow students to focus only on materials selection. By doing so, the focal point of the activity becomes the insulation of materials, not structure design. The scenario can be altered and thus shift the problem from insulation that keeps heat from transferring out to insulation that protects from heat transferring in. This activity provides students with active experiences as they explore heat and thermal insulation, experimental design and trials, and revisions. There are numerous adaptations of scenarios and materials, but the main goal is to create a classroom environment that provides students with problem-solving opportunities that require them to think creatively, explore, sometimes fail, and if necessary, try, try again.

References

Attenborough, D. 1979. *Life on Earth*. Boston, MA: Little, Brown and Company.

Bybee, R. W. 1997. *Achieving scientific literacy: From purposes* to practices. Portsmouth, NH: Heinemann.

Paik, S. H., B. K. Cho, and Y. M. Go. 2007. Korean 4- to 11-year old student conceptions of heat and temperature. *Journal of Research in Science Teaching* 44 (2): 284–302.

Resource

Royston, A. 2001. *My world of science: Hot and cold*. Chicago, IL: Heinemann Library.

Internet Resources

Chem4kids: Heat and Cold
    www.chem4kids.com/files/react_thermo.html
Department of Energy—Insulation Fact Sheet
    www.ornl.gov/sci/roofs+walls/insulation/ins_06.html

CHAPTER 24

Nanoscale in Perspective

By Elvis H. Cherry, Weijie Lu, and R. P. H. Chang ///

Most students have an understandably hard time imagining something as small as a nanometer: 1,000,000,000 of a meter. However, nanoscale science is a growing field, and to appreciate the work of scientists in this field, it is important for students to understand the scale of work being done. To start, give students the following quiz:

Which of the following statements are fact, and which are fiction?

A. Robots can be injected into your bloodstream to repair damaged organs.

B. A tiny guitar the size of a red blood cell has been constructed.

C. Electricity can be produced from the chlorophyll of spinach.

Currently, A is still a work in progress, but B and C have already been accomplished. Nature has provided researchers with the inspiration for bloodstream robots, the details of which can be found at *www.kurzweilai.net/ meme/frame.html?main=/articles/art0410.html?*. The 10-micrometer nanoguitar with six strings that are 50 nanometers wide was created by Cornell University (*www.oddmusic.com/gallery/om22000.html*). Researchers at Massey and Vanderbilt Universities have created

---

FIGURE 24.1.

Size scale
Proton = 1 x 10^{-15} m
Hydrogen atom = 10^{-10} m
Buckyball (C60) = 10^{-9} m
Virus = 10^{-8} m
Bacterium = 10^{-7} m
Cell nucleus = 10^{-6} m
Red blood cell = 10^{-5} m

---

chlorophyll-based photosensitive film that is one nanometer thick and capable of generating electricity. Grab students' interest with these innovations and follow up with this nanotech activity created by the National Center for Learning and Teaching in Nanoscale Science and Engineering (NCLT). Before starting, your students should be familiar with the powers of 10 and the metric system.

This activity, designed to bring nanoscale into the familiar macroworld, is from a two-week workshop on nanotechnology conducted by Fisk University in Nashville, Tennessee that was sponsored by NCLT and the National Science Foundation. Other nanoscale activities are available at NCLT's website at *www.nclt.us/ nclt.html*.

Acknowledgment

This work is supported by the National Center for Learning and Teaching in Nanoscale Science and Engineering (NCLT) under the National Science Foundation Grant # 0426328.

Activity: Powers of 10

Materials
- one roll of register tape
- scissors (students should exercise caution when handling scissors)
- meterstick
- adhesive (transparent) tape

Procedure
Part 1

- Cut the register tape into strips of paper that are 1 mm, 1 cm, 1 dm, 1 m, and 1 decameter in length. (More than one group can create the same length tape, if needed.)
- In the hallway, or other open space, tape the strips side by side.
- Label each strip in meters using decimal points and the powers of 10. For example, 1 mm is 0.001 meters, or 10^{-3} m.
- Explain what happens each time you move the decimal point or increase/ decrease the exponent.

Part 2

- Add new labels to the strips to represent the powers of ten at the nanoscale. To the 1 mm strip, add the label 10^{-9} m, to the 1 cm add 10^{-8} m, and so on.
- Select items from Figure 1 that correspond to the nanoscale measurements.

CHAPTER 25

What's So Big About Being Small?

By MaryKay Orgill and Kent J. Crippen //

The term *interdisciplinary* implies a blending of two perspectives while continuing to recognize each as unique and distinct. It should not be confused with the term *integration*, which denotes the blending of two or more ideas into one. This is an important distinction. According to Lederman and Ness (1997):

> In an interdisciplinary curriculum/instructional approach, the integrity of the various academic disciplines remains clear. No attempt is made to 'blur' the distinctions between and within mathematics and the sciences. Although the connections between subject matters are emphasized, there remains perceived value in the unique characteristics and distinctions among the various disciplines. (p. 2)

An interdisciplinary approach to teaching involves leveraging the different perspectives of each discipline to better understand an issue or problem. Researchers suggest that this approach enhances meaningful learning (Lonning and DeFranco 1994, 1997) and that it more closely approximates real-world learning and problem solving than a discipline-specific method (Everett 1992). However, despite the potential benefits, not all topics lend themselves to an interdisciplinary approach. The most ideal topics for interdisciplin-

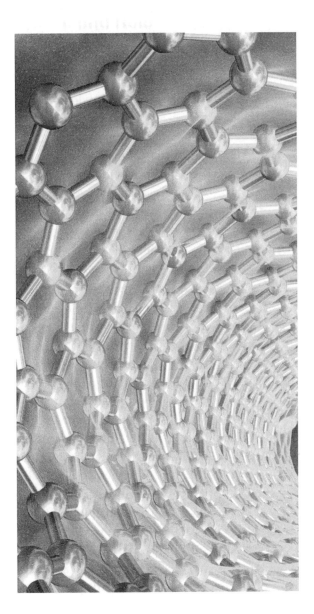

ary study are those whose very nature is also interdisciplinary. Nanoscience—which combines biology, chemistry, physics, engineering, and mathematics—is one such topic.

This article presents the use of nanoscience as a theme for the interdisciplinary study of science and mathematics. We present an example curriculum and thoughts about an inquiry instructional method. The activities in this curriculum are appropriate for both middle and high school students.

Interdisciplinary Nanoscience

Nanoscience involves the investigation of matter that exists in the size range of 1–100 nanometers (1 nm = 10^{-9} m). The behavior of most objects we experience every day can be described with classical mechanics. However, objects at the nanoscale (and smaller) exhibit unique, nonclassical properties. For example, at the nanoscale, copper is transparent, carbon can conduct electricity, aluminum is combustible, and gold can be red in color. These unique properties, which can only be explained with quantum mechanics, make nanoscale materials interesting to scientists.

As a research field, nanoscience requires an interdisciplinary perspective and is forcing many researchers to rethink the nature of their work and leverage collaborative efforts to provide needed interdisciplinary expertise. For example, materials development at the nanoscale involves the manipulation of matter and energy (traditionally the domain of engineering), requires knowledge of atomic and subatomic properties (traditionally the domains of chemistry and physics, respectively), and is often applied within living systems (traditionally the domain of biology). Because nanoscale objects cannot be directly experienced using the senses, understanding and examining them often involves modeling (traditionally the

domain of both science and mathematics).

As content, nanoscience offers a unique opportunity for interdisciplinary teaching and learning. Several teachers have previously suggested that nanoscience can be used as a context for teaching either mathematics or science (Jones et al. 2007; Stevens et al. 2007; Tretter 2006), and we argue that approaching this content from an interdisciplinary perspective can lead to a better understanding not only of an exciting developing field, but also of underlying mathematics and science concepts.

Big Ideas About Little Things

The National Center for Learning and Teaching in Nanoscale Science and Engineering (NCLT) has been instrumental in facilitating discussions between researchers and science educators about the key scientific ideas that set nanoscience apart as a unique discipline. Through a series of national workshops, the NCLT identified eight "Big Ideas of Nanoscience" appropriate for study at the middle and high school levels:

- Size and Scale
- Structure of Matter
- Size-Dependent Properties
- Forces
- Self-Assembly
- Tools and Instrumentation
- Models and Simulations
- Nano and Society

To develop an interdisciplinary nanoscience curriculum, it was essential to identify key mathematics and science concepts that underlie and relate to these Big Ideas. Our goal was to develop an interdisciplinary curriculum based on

FIGURE 25.1.

Concepts from the science and mathematics standards (NRC 1996; NCTM 2005).

Big Ideas	Science focus	Mathematics focus	Inquiry question
Size and Scale	Measurement Precision and accuracy Estimation	Proportional reasoning Big and small numbers Error	How small is small?
Structure of Matter; Forces	Structure of matter Intermolecular forces	Graphing	How are small things arranged?
Size-Dependent Properties	Size-dependent properties Experimental design Properties of water	Surface area and volume Ratio and proportion	What are the rules in a small world?
Self-Assembly	Self-assembly Collision theory Intermolecular forces Cells	Probability Surface area and volume Ratio and proportion	How can simple rules create complex things?
Tools and Instrumentation	Graphing Modeling Appropriate tools Experimentation Data collection	Integers Graphing Absolute value Measurement Scaling Two-dimensional and three-dimensional graphs	How do we build small things?
Nano and Society	Nature of science	Number sense Ratio and proportion	Nano and Society: What's the big deal?
Models and Simulations	Models Visualization Analogies	Forms of representation Prediction	No inquiry question: Overarching theme for all "Big Ideas"

nanoscience that includes mathematics as well as the traditional science disciplines of biology, chemistry, and physics. While many of the Big Ideas correlate well with concepts in the current *National Science Education Standards (NSES)* (NRC 1996), the connections to mathematical concepts were not immediately obvious to us. Through consultation with colleagues in mathematics education and in the local school district, we identified key underlying mathematical concepts for each of the Big Ideas from the National Council of Teachers of Mathematics (NCTM) Principles and Standards (the mathematics equivalent of the NSES) (2005). The result was a matrix of interrelated science and mathematics ideas grounded in the theme of nanoscience through the Big Ideas (Figure 25.1, p. 171).

For each of the Big Ideas, we created an inquiry question that could be used to drive a smaller portion of an interdisciplinary unit about nanoscience. The related Big Ideas of the Structure of Matter and Forces are grouped into the same subunit. There is not an inquiry question for Models and Simulations, however, because we felt that this Big Idea should be an overarching theme for all the others. For example, in order to understand the properties of nanoscale objects, we often use physical models such as Lego blocks or interactive computer simulations. Although the models approximate the smaller objects, we have to take into account the assumptions used in building the models and their inherent limitations, as well as the fact that objects at the bulkscale behave differently than objects at the nanoscale.

An Interdisciplinary Nanoscience Curriculum

A nanoscience unit can be interdisciplinary in both its content and its method of presentation. Inquiry, as a method of knowing and conceptual-

izing the physical world, is the glue that binds the theme of nanoscience. The five essential features of inquiry found in the NSES (NRC 2000) align well and complement those in mathematics (NCTM 2005; Figure 25.2).

Because of this, the framework for our interdisciplinary unit on nanoscience is grounded in both the five essential features of inquiry (NRC 2000) and the 5E instructional model (Bybee 1993). Our framework includes inquiry questions, topic questions, activities, and closing questions for each of the Big Ideas. Inquiry questions are used to motivate and engage students. Opening topic questions break down the bigger themes into manageable concepts and focus the inquiry while allowing students to express their prior knowledge about each of the Big Ideas. Some activities allow students to explore the Big Ideas and various nanotechnology products, while others emphasize the process of developing knowledge claims from evidence and generating multiple representations of data with technology to foster conceptual change. Closing questions redefine the concepts and connect the bigger themes into a larger framework (Figure 25.3, pp. 174–175).

There are many excellent resources for nanoscience activities. In fact, the majority of the activities listed in Figure 25.3 came from the internet or from NSTA and include their own assessment materials (see References). Our contribution was to select, sequence, and organize a set of activities into a unique, interdisciplinary nanoscience curriculum. The rows in Figure 25.3 can be thought of as individual subunits, each of which can be used as either part of a larger collection or separately, as long as the teacher recognizes that some of the ideas presented earlier in the unit are necessary for understanding ideas presented later. The following section describes our implementation of the subunit on Size-Dependent Properties to illustrate the unit's interdisciplinary focus with a detailed example.

Size-Dependent Properties Subunit

In the third subunit, a series of activities explores the inquiry question "What are the rules in a small world?" from physical, mathematical, and chemical perspectives in order to help students learn more about the Big Idea of Size-Dependent Properties. The 5E cycle is used in each of the activities. Investigations are initiated with a framing question (engagement), which is followed by data collection (exploration). The data is then interpreted from both a scientific as well as a mathematical perspective (explanation), generalized to other conditions such as the atomic- or bulkscale (elaboration), and assessed for mastery (evaluation).

By the time students start this subunit, they have discussed and completed activities about absolute and relative scale, measuring and studying objects that are either too big or too small for us to sense directly, the structure of the three physical states of matter, and the relationships between chemical structure and the strength of intermolecular forces. Students have also explored the unique properties of different nanotechnology products.

Throughout the previous subunits, it is emphasized that nanoscale materials behave differently than materials on the bulkscale, and this subunit allows for an exploration of why. To begin the subunit, students apply what they have learned about liquids and intermolecular forces in a previous subunit by responding individually to the following question: "How does the size of a container affect the ability of liquid to flow from it?"

FIGURE 25.2.

An interdisciplinary perspective on inquiry

Essential features of inquiry found in the NSES (NRC 1996)	Statements from the NCTM standards (2005)
Engagement in scientifically oriented questions occurs	Formulate questions that can be addressed with data and collect, organize, and display relevant data to answer them (Data analysis and Probability)
Priority is given to evidence in responding to questions	Make and investigate mathematical conjectures (Reasoning and Proof)
Explanations are formulated from evidence	Create and use representations to organize, record, and communicate mathematical ideas (Representation)
Explanations connect to scientific knowledge	Build new mathematical knowledge through problem solving (Problem solving)
Explanations are communicated and justified	Organize and consolidate mathematical thinking though communication (Communication)

Integrating Engineering + Science in Your Classroom

FIGURE 25.3.

Nanoscience curriculum map: Connecting science and math

(**Note:** See References for links to these activities.)

Inquiry question	Opening topic questions	Activities	Closing topic questions
How small is small?	What is the biggest measurable object you can think of? What is the smallest measurable object you can think of?	Size and Scale Card Sort (NCLT 2006c) One in a Billion (Jones et al. 2007)	List one example object for each of the following sizes: 10^3 m, 10^0 m, 10^{-3} m, 10^{-6} m, 10^{-9} m. Explain the advantages and disadvantages of using the human senses to detect things in small quantities. Describe other alternatives.
How are small things arranged?	Why do you think some substances are solid at room temperature while others are liquids or gases? What is the value of different types of graphical representation of data?	Nanotech, Inc. (Jones et al. 2007) Structure of Matter and Heating Curves (NCLT 2006d) Intermolecular Attractions (Holmquist, Randall, and Volz 2007)	Explain the following statement: If you were very tiny, everyday activities such as eating, drinking, and maintaining temperature would require a different set of challenges than those of the world you are used to inhabiting.
What are the rules in a small world?	How does the size of a container affect the ability of liquid to flow from it?	Putting Nano-Tex to the Test (Kao, Zenner, and Gimm 2005) Physics Changes With Scale (Jones et al. 2007) Shrinking Cups (Jones et al. 2007) Viscosity of Liquids (Jacobsen and Moore 1998) Computer Activity: Intermolecular Forces (Molecular Literacy Project 2008)	Explain the nature and relative strength of the possible forces between molecules. Use your experience with these activities to explain why spiders cannot be 3 m tall.

FIGURE 25.3. (continued)

Inquiry question	Opening topic questions	Activities	Closing topic questions
How can simple rules create complex things?	Which factors do you think will influence whether two molecules will react together in a test tube? Which factors do you think will influence whether two molecules will react together in a living system?	Limits to Size: Could King Kong Exist? (Jones et al. 2007) Computer Activity: Self-Assembly (Molecular Literacy Project 2008) Self-Assembly of Legos (NCLT 2006a; Jones et al. 2007)	How does the surface area of an object change as it gets infinitely bigger or smaller in volume? What impact would nonlinear changes in surface area to volume have for the properties of objects that get bigger or smaller? Explain how the laws of probability impact self-assembly.
How do we build small things?	There are many objects that are either much smaller or much larger than those we directly experience in our everyday lives. What types of mathematical and scientific tools or instrumentation might help us develop models to better understand the way these objects work?	Nanomaterials: Memory Wire (Jones et al. 2007) What's in Your Bag? Investigating the Unknown (Jones et al. 2007) Probe Microscope Model: Refrigerator Magnet (MRSEC 2005) Video: *Probe Microscopes* (MRSEC 2007) Scanning Probe Microscopy (NCLT 2006b; Jones et al. 2007) Computer Activity: Graphing in Excel (Bellomo 2008)	Reflect on the tools and instrumentation used to help us better understand our world. Identify two you are most comfortable using and two you learned more about in this unit. What role does scale play when using tools and instrumentation to understand our world?
Nano and Society: What's the big deal?	Describe some potential threats as well as benefits that might emerge from the use of advances in nanotechnology.	Video: *Power of Small* (Fred Friendly Seminars and ICAN Productions 2008) Too Little Privacy: Ethics of Nanotechnology (Jones et al. 2007) Reading: "Nanotechnology—Untold promise, untold risk" (CUUS 2007)	Have your ideas about nanoscience and nanotechnology changed as a result of participating in this unit? Have any specific concepts or ideas in this unit sparked your interest? Explain.

Putting Nano-Tex to the Test

The first activity in this subunit allows students to explore a product of nanotechnology: Nano-Tex stain-resistant fabric, which is now found in many clothing products. In this activity, Putting Nano-Tex to the Test (Kao, Zenner, and Gimm 2005), students treat a variety of fabric swatches—including a swatch of Nano-Tex—with different staining agents to determine which fabric is most stain-resistant and whether certain fabrics are more resistant to particular types of stains. Through this and previous activities, students can see that products made of or treated with nanoscale materials behave differently. The question is "Why?"

To further explore this question, students complete two additional activities from the *Nanoscale Science* book (Jones et al. 2007): Physics Changes With Scale and Shrinking Cups: Changes in the Behavior of Materials at the Nanoscale. Because nanoscale objects cannot be directly seen or manipulated in a typical science classroom, both of these activities use models or analogies to illustrate concepts that are important at the nanoscale.

Physics Changes With Scale

One concept we want students to understand is that while there are some properties that are independent of an object's size, other properties are dependent on how big the object is. Teachers can help students explore this concept with a physics analogy about inertia and drag forces that also involves the mathematical concepts of surface area and volume. Specifically, we look at the competition between inertia (resistance to change of velocity) and drag forces (air resistance) and how the dominance of one over the other changes with the size of an object.

Consider what happens when students are asked to measure the distance they can throw Styrofoam balls ranging in size from 1.3 to 10 cm in diameter down a hallway. Students quickly discover that they can throw the larger balls farther than the smaller balls, so we ask them to collect data about the balls to determine which properties affect the distance they can be thrown. Students measure the mass and diameter of the balls and then calculate the surface area, volume, and density of each.

As a result, students determine that density does not change with the size of the balls, but other properties—such as the mass, volume, and surface area—do change. Teachers can use these results to reinforce the concept that density is an intrinsic property of a substance before asking students which of the other properties seem to affect the distance the ball can be thrown. Students may indicate that according to the data they have collected, surface area seems to be the most important factor, but they may not know why this is the case.

Students' findings lead nicely into a conversation about the physics concepts of drag forces and inertia. Briefly, inertia is directly proportional to the mass of an object (and by extension, volume, for objects having the same density), while drag forces are directly proportional to an object's surface area. For two similarly shaped but different-size objects traveling at the same velocity in the same medium, deceleration (and thus the distance the object can travel) is affected by the ratio of drag forces to inertia and, proportionally, to the ratio of surface area to volume. For larger objects, inertia (which is related to volume) dominates. (**Note:** Although inertia is always related to mass, in this case it is related to volume as well, since the Styrofoam spheres all have the same density, as students have discovered.) For smaller objects, drag (which is related to surface area) dominates.

This change in the dominance of one property over another is analogous to a change in dominance of forces that affect macroscale versus nanoscale objects. At the macroscale, gravitational forces, which depend on the mass of an object and the distance between two objects, dominate. For low-mass nanoscale objects, intermolecular interactions—which depend on the detailed distribution of charge, polarizability, and various other quantum-scale effects—dominate.

This physics analogy for nanoscale concepts sets the stage for a discussion of the mathematical concepts of surface area and volume calculations and how they change disproportionately as objects become smaller. This knowledge is critical for understanding why nanoscale materials behave differently than bulkscale materials. According to Jones et al., "At the nanoscale, almost all interactions are mediated by surface effects. So forces between objects are often proportional to their surface area" (2007, p. 84). Surface-related interactions, such as intermolecular attractions, become very important at the nanoscale, where the surface-area-to-volume ratios (SA/V) are large. Take, for example, a comparison between a macroscale sphere with a 1 cm radius and a nanoscale sphere with a 1 nm radius. The macroscale sphere has an SA/V ratio of 3×10^2/m, but the nanoscale sphere has an SA/V ratio of 3×10^9/m—10 million times bigger than that of the macroscale sphere!

Shrinking Cups

The next activity—Shrinking Cups—reinforces the idea that different forces dominate at different scales, while allowing the class to revisit concepts about intermolecular forces. Students receive two cups: a normal drinking cup and a small, dollhouse-size cup. They know that if a normal-size cup is tipped, water will pour out, so they predict what would happen if the small, dollhouse-size cup were filled partway with water and then tipped. When testing their predictions, many students may be surprised to find out that the water does not pour out of the small cup.

To explore this phenomenon further, students create cups made of clay in different shapes and sizes. Students should count how many drops of water their cups can hold without the water pouring out when it is tipped. In general, the groups whose containers hold the most water when tipped create the cups with the smallest openings. In most cases, though, there is a point at which adding one more drop of water will cause all the water to pour out when the cup is tipped.

After students create their cups, they attempt to explain why water does not initially pour out of their small cups. Often they are not sure, so to get them thinking, teachers can ask what students know about water. Based on activities from previous subunits, students may talk about the strong intermolecular forces (hydrogen bonding) in water, and teachers can extend their thoughts into a discussion of water's properties typically covered in a chemistry class (e.g., polarity, adhesion, cohesion, and surface tension).

Students then think about why there is a point at which the addition of one drop of water could cause all of the water to tip out. It usually takes very little prodding—and a reminder of the previous activity—for students to come to the conclusion that, although intermolecular forces dominate for small objects, at a certain point, the addition of a drop of water would cause gravitational forces to overcome the intermolecular forces.

Students then predict what would happen if they used a different liquid—such as cooking oil or corn syrup—in their small containers and conduct experiments. They find that some liquids (e.g., oil) pour more easily from their cups than others (e.g., corn syrup). Based on previous discussions, students can determine that the ability of a liquid to flow is influenced by its intermolecular forces: the stronger the intermolecular forces, the slower the liquid flows. At this point, students have discovered one of the main reasons for the different viscosities of different liquids. It is even more exciting when students ask to see the chemical structures of the liquids' components so they can rationalize the difference in the relative strengths of intermolecular forces.

Viscosity of Liquids

Next, students extend what they have learned from the Shrinking Cups activity and integrate this knowledge with information previously learned about intermolecular forces. As an interactive group activity, students are introduced to the chemical structures of different liquids and then predict which liquid will flow more rapidly. After making their predictions, students can watch the *Viscosity of Liquids* videos from the *Journal of Chemical Education*'s *Chemistry Comes Alive!* series (Jacobsen and Moore 1998), which show the liquids flowing from pipettes (see Internet Resources). In most cases, students can correctly predict which liquid will flow the fastest and explain why this is so based on structural features and the corresponding intermolecular forces.

At the end of the subunit, students respond to the following: "Explain the nature and relative strength of the possible forces between molecules" and, separately, "Use your experience with these activities to explain why spiders cannot be 3 m tall." In response to the first probe, students can often accurately describe structural features of molecules and how those features affect the relative strength of intermolecular forces. Their answers to the second probe are typically more varied. Students sometimes have difficulty connecting the information from this subunit to a biological situation; however, we believe that responding to this question prepares them for future subunits and activities (e.g., examining why the size of a cell is affected by surface-area-to-volume ratios using the Limits to Size: Could King Kong Exist? activity from *Nanoscale Science* [Jones et al. 2007]).

Although this subunit about size-dependent properties is only a small part of the overall nanoscience unit, it is interdisciplinary in nature and accomplishes several learning goals—all in the context of the new and exciting field of nanoscience:

- It reinforces previous learning about the structure of matter and intermolecular forces and connects this knowledge to the mathematical concepts of surface area and volume.

- It allows teachers to introduce physics concepts of inertia and drag and chemistry concepts of polarity, adhesion, cohesion, and surface tension.

- It prepares students for the next subunit, in which teachers discuss some of the biological implications of surface-area-to-volume ratios.

Conclusion

As a discipline, nanoscience offers a unique opportunity for interdisciplinary study. Nanoscience has developed a mystique in today's pop culture, and futurists speak lavishly about the potential of nanomaterials. Students are motivated by the Big Ideas of nanoscience and are intrigued to better understand nano-engineered products—such as stain-resistant pants and shirts, socks that prevent foot odor, or shirts that can roll up their own sleeves in warm weather—many of which are already available commercially. When interdisciplinary nanoscience activities are framed from an inquiry question, student motivation for the Big Ideas easily translates into deeper and sustained learning of basic mathematical and scientific principles that emanate from students' bulk-world experiences. Using multiple perspectives to interpret experimental data fulfills the vision of interdisciplinary instruction by emphasizing key relationships among mathematics and science disciplines while valuing the unique characteristics and distinctions of each. Making the important connection between basic principles and the application to commercial products and techniques affords students the opportunity to better understand the world in which we live.

Acknowledgments

We would like to thank Carryn Bellomo (University of Nevada–Las Vegas Department of Mathematical Sciences), Jeff Bostic (Southern Nevada Regional Professional Development Program), and the project facilitators of the Clark County School District's Curriculum and Professional Development Division—Kerrie Blazek, Jodi Cunningham, and Tina Mika—for their collaboration in developing and implementing this unit. Funding for this project was provided by a state Mathematics and Science Partnership grant.

References

Bellomo, C. 2008. *Computer activity: Graphing in Excel.* Las Vegas: University of Nevada.

Bybee, R. 1993. An instructional model for science education. In *Developing biological literacy*. Colorado Springs, CO: Biological Sciences Curriculum Study.

Consumers Union of United States (CUUS). 2007. Nanotechnology: Untold promise, untold risk. *Consumer Reports* July: 40–45.

Everett, M. 1992. Developmental interdisciplinary schools for the 21st century. *Education Digest* 57: 57–59.

Fred Friendly Seminars and ICAN Productions. 2008. Nanotechnology: The power of small. *http://powerofsmall.org*

Holmquist, D. D., J. Randall, and D. L. Volz. 2007. Evaporation and intermolecular forces. In *Chemistry with Vernier,* pp. 9-1–9-4T. Beaverton, OR: Vernier Software and Technology.

Jacobsen, J. J., and J. W. Moore. 1998. Viscosity of liquids. In *Chemistry comes alive!*, Volume 2 (CD-ROM). Madison, WI: *Journal of Chemical Education*.

Jones, M. G., M. R. Falvo, A. R. Taylor, and B. P. Broadwell. 2007. *Nanoscale science: Activities for grades 6–12.* Arlington, VA: NSTA Press.

Kao, Y. S., G. M. Zenner, and J. A. Gimm. 2005. Putting Nano-Text to the test. *Science Scope* 29(6): 37–41.

Lederman, N. G., and M. L. Ness. 1997. Integrated, interdisciplinary, or thematic instruction? Is this a question or is it questionable semantics? *School Science and Mathematics* 97(2): 57–58.

Lonning, R. A., and T. C. DeFranco. 1994. Development and implementation of an integrated mathematics/science preservice elementary methods course. *School Science and Mathematics* 97: 18–25.

Lonning, R. A., and T. C. DeFranco. 1997. Integration of science and mathematics: A theoretical model. *School Science and Mathematics* 97: 212–215.

Materials Research Science and Engineering Center on Nanostructured Interfaces (MRSEC). 2005. Refrigerator magnet activity guide. University of Wisconsin. *www.mrsec.wisc.edu/Edetc/supplies/ActivityGuides/Refrig_Magnet_Guide_2005.pdf*

Materials Research Science and Engineering Center on Nanostructured Interfaces (MRSEC). 2007. Probe microscopes: Tools used in nanotechnology. University of Wisconsin. *www.*

mrsec.wisc.edu/Edetc/cineplex/nanoquest/tools.html

Molecular Literacy Project. 2008. Concord consortium. *http://molit.concord.org*

National Center for Learning and Teaching in Nanoscale Science and Engineering (NCLT). 2006a. How is a self-assembling system designed? *www.nanoed.org/nlr/Purdue/Self-Assembly%20-%20How%20is%20a%20Self-Assembling%20System%20Designed.pdf*

National Center for Learning and Teaching in Nanoscale Science and Engineering (NCLT). 2006b. Scanning probe microscopy lesson. *www.nanoed.org/nlr/Purdue/Scanning%20Probe%20Microscopy.pdf*

National Center for Learning and Teaching in Nanoscale Science and Engineering (NCLT). 2006c. Size and scale: A card sort. *www.nanoed.org/nlr/Purdue/Size%20and%20Scale%20-%20A%20Card%20Sort%20Activity.pdf*

National Center for Learning and Teaching in Nanoscale Science and Engineering (NCLT). 2006d. The structure of matter. *www.nanoed.org/nlr/Purdue/Structure%20of%20Matter.pdf*

National Council of Teachers of Mathematics (NCTM). 2005. Principles and standards for school mathematics. *http://standards.nctm.org*

National Research Council (NRC). 1996. *National science education standards.* Washington, DC: National Academies Press.

National Research Council (NRC). 2000. *Inquiry and the national science education standards: A guide for teaching and learning.* Washington, DC: National Academies Press.

Stevens, S., N. Shin, C. Delgado, C. Cahill, M. Yunker, and J. Krajcik. 2007. Fostering students' understanding of interdisciplinary science in a summer science camp. *www.nanoed.org/nlr/fostering_students_understanding/UM-stevens-SSI_NARST_07.pdf*

Tretter, T. 2006. Conceptualizing nanoscale. *The Science Teacher* 73 (9): 50–53.

Internet Resources

Videos and still images for *Viscosity of Liquids* activity

http://jchemed.chem.wisc.edu/jcesoft/cca/CCA2/MAIN/VISCLIQ/CD2R1.HTM and *http://jchemed.chem.wisc.edu/jcesoft/cca/CCA2/SMHTM/VISCLIQ.HTM*

PART THREE

After-School Programs

CHAPTER 26

Fueling Interest in Science
An After-School Program Model That Works

By Kathleen Koenig and Margaret Hanson ///

G irls in Science (GIS), a unique after-school program for sixth- and seventh-grade girls, was created by a group of educators concerned with increasing the number of girls interested in science, technology, engineering, and math (commonly referred to as the STEM disciplines). The group recognized the poor representation of women in STEM careers and the research that indicates girls tend to lose interest in science and math in the middle school years (Catsambis 1995; Farenga and Joyce 1999; Jones, Howe, and Rua 2000; Blue and Gann 2008). In addition, several in the group had firsthand experience with college-age women who had severely limited their choices of college majors by opting out of upper-level math and science courses in high school.

The GIS planning group was passionate about designing a program for middle school girls that would stimulate and continue their interest in science and math. After a year of research and planning, the group developed a program model that has remained essentially unchanged since its implementation during the 2002–03 school year. Program goals include (1) educating about science and math in everyday applications with an emphasis on STEM careers, (2) providing opportunities for girls to experience success in science and math, (3) challenging girls to continue in science and math with a plan for high school

and beyond, and (4) educating parents on how to support daughters interested in STEM.

The program model includes roughly eight monthly after-school meetings with women scientists, one Saturday field trip to a local university, and an end-of-year poster session in which each girl in the program does a mini-presentation on a science career. Mothers of the participating girls are invited to all program events as a way to educate them on the vast domain of STEM careers and education opportunities available to their daughters. Only female role models have been encouraged to attend the program events as a means of keeping the environment single gender. Teachers involved in the program have observed that girls participate more in female-only environments, a finding that is also supported in the research literature (Sax 2006).

Monthly After-School Meetings

The monthly after-school meetings are scheduled by the designated school GIS coordinator (typically the science teacher) after consulting a list of approximately 15 women scientist volunteers posted on the program website. During these one-hour meetings held on individual school grounds with typically 15–30 girls in attendance, the invited scientist presents her career field, provides a related hands-on activity, and discusses the education and experience necessary for her current position. She may share other aspects of her career, including how her job is helpful to others, the opportunities that exist for women in her field, the difficulties she has encountered, the type of support she has received from family and friends, and if relevant, what it is like to be a wife or mother and have a career in STEM. In essence, the women scientists become role models for the girls by demonstrating that women in STEM careers are real people with real lives and interests.

The scientist presenters in GIS represent a range of disciplines (physics, chemistry, biology, geology, engineering, math, and technology) and a range of degrees (associate's, bachelor's, master's, and doctoral). They are recruited yearly via e-mails sent to local universities and industries

or by word of mouth. Most continue to volunteer for the program beyond their first year of involvement. Some of the activities these women have provided include taking apart toasters and curling irons for circuit analysis, investigating the effect of fresh pineapple in Jell-O, testing classroom surfaces for microbes, examining fossils to extract information about extinct species, analyzing soil as it relates to storm-water retention design, and exploring music through computers.

Showcase of Women Scientists

One Saturday each spring, all girls participating in GIS are invited to a local university for a showcase of women scientists. The event provides the girls with experiences in the actual research labs of women scientists working at the university, while educating potential college STEM majors. The event includes women scientists who do not participate in the monthly meetings at the schools, so the topics and activities are new to all who attend. The girls spend approximately 50 minutes in each of four labs and participate in related activities. Past experiences have included analysis of pond water, impact of aerodynamics, chromatography, website creation, production of nylon in a test tube, and investigation of construction and architectural design.

End-of-Year Poster Presentation

At the final after-school meeting, each student presents a poster titled "If I were a scientist, I would be a..." Creating the poster encourages the girls to reflect back on the year and focus on a single STEM career of interest. Elements of the poster include why this career was chosen, a description of the activities a scientist in this field would engage in and how her work would be useful to society, a list of high school science and math courses essential for this career, and a list of activities the girls could engage in now to learn more about this field of study. Women scientists who worked with the girls at previous GIS meetings are invited to the poster session to offer additional insight and support.

Program Success

GIS has quickly become popular and more schools adopt the program each year. Two parallel programs are now offered in Cincinnati and Dayton, Ohio, and combined there are 22 schools currently participating. The retention of schools participating in the program has been encouraging, with 22 of the 27 schools that adopted the program between 2002 and 2008 still continuing to offer it. The five schools that dropped the program did so only after the school program coordinator was no longer teaching at the middle school grade level or was reassigned to a different school.

The program has grown significantly since it was first piloted in 2002. It is estimated that over 1,600 students have now participated in GIS. Attendance records indicate that the program is very popular within the schools themselves. Many of the program coordinators report that 90% or more of their sixth- and/or seventh-grade girls are involved in the program, with average attendance rates for monthly meetings typically above 80%. Teachers were surprised by the large number of girls who regularly attended the monthly meetings.

A survey of the school program coordinators has provided very positive feedback about the program. In particular, the coordinators like (1) the ease of using the GIS website to choose and schedule scientist presenters each month, (2) the quality of presentations by the scientists, and (3) the overall program model, with monthly meet-

ings held directly after school on school grounds. A representative sample of comments include the following:

- "The girls looked forward to the events. They didn't seem inhibited at these meetings like they sometimes do in class. They were not fearful about asking or answering questions."

- "The girls seem more interested in careers and the mothers seem more interested in their daughters' academics and interest in science."

- "The girls have a very positive opinion of science. I think that the program has helped with their knowledge and attitudes toward science."

Mothers of the girls participating in the program were also surveyed and most responded positively to questions concerning (1) the time and location of program events, (2) the range of STEM careers presented, (3) the observed interest of the girls in attending the events in addition to discussing events at home, and (4) the nature of the events themselves. Mothers particularly liked the hands-on activities, the girls-only environment, and that mothers were invited to all program events. Although most indicated no changes were needed to the program, several working mothers requested that meetings be held in the evenings so they could attend with their daughters. Others requested meetings be held more often than once a month or suggested additional careers be represented, such as doctors or veterinarians. A representative sampling of comments included the following:

- "My daughter's interest in science or future science careers has grown."

- "I learned more about women in science than I thought I would."

- "I liked that the moms were included in the events. After attending, my daughter and I had a lot to talk about regarding the activities and careers in science."

- "The program is an excellent opportunity for young girls to explore the field of science and its opportunities."

- "We both learned what a huge field of different interest is available."

- "My daughter never talked about a career in science until this program."

- "This was a great program. Bringing science out of textbooks and into its practical applications is a great concept for girls this age."

Although we were delighted that the program model was found to be popular with the girls, coordinators, and parents, our ultimate goal for the program is to positively impact girls' attitudes toward and interest in STEM and STEM careers. Changes in attitude and interest were assessed through Likert-scale surveys administered to the girls at the beginning and end of their GIS experience. The survey questions were validated by a panel of science education experts. Ninety percent of the girls who participated in GIS between 2002 and 2006 had matching pre- and postquestionnaires and significant positive changes were seen on the following statements: (1) I would enjoy having a science career when I am older, (2) Science is boring, (3) Science classes provide me with skills to use outside of school, (4) Anyone interested in science can become a scientist, (5) Being a scientist would be fun, (6) My friends would approve if I chose a career in

science, and (7) I know, or have talked to, women scientists who are really cool.

Program Implementation

A similar program can be implemented in any school or region and specific details of the steps involved can be found on our GIS website, linked through *www.wright.edu/~kathy.koenig*. The website also includes sample letters, e-mails, and forms we use each year in recruiting the participating schools and women scientist volunteers. For convenience, a brief summary of program implementation steps is included here. First, women scientist presenters must be recruited, and we have found that e-mails sent through contacts at local universities, industries, and public service organizations work well. These e-mails should provide details about the role and time commitment of the presenter and a form should be included that can be filled out and returned by those interested in volunteering. Second, the presenter contact information and presentation description gathered from the returned forms should either be uploaded to a program website or put into a program booklet. Participating schools will need access to this information to choose and schedule presenters throughout the academic year. Third, participating schools must be recruited, and we have found that invitation letters sent directly to all sixth- and seventh- grade science teachers within the target region work well. The letters should contain specific program details, a registration form, and either a program website address or contact information of the program organizer for those interested in learning more.

Once a school returns a registration form, either a paper copy of the list of scientist presenters or a link to the online list of presenters should be sent to the school program coordinator. We do not send out this information until a school registers so we can better monitor the program.

Fourth, once the program is up and running, all participating school program coordinators should be contacted bimonthly to determine which presenters are being invited to the schools—whether or not the presentations support the program goals—and to offer any needed support. And finally, the Showcase of Women Scientists event held at a local university must be planned. Although we both work in university science departments and have found this task manageable, this may not be the case for other program organizers. Because the showcase is not an integral part of the academic year program, it may be omitted. See our GIS website for more information.

Discussion

As our society becomes more technologically advanced and jobs require additional related skills, it is important that all girls, not just those interested in STEM, take advanced levels of science and math in high school. Evidence suggests that intervention programs such as GIS can improve girls' interest in and attitude toward science, which may in turn impact future course selections and career choices (Gaston 2001; McCormick and Wolf 1993; Stake and Mares 2001). In light of this, we are encouraged that GIS has had a demonstrated impact on girls' attitudes toward STEM and has had such positive responses from those involved in the program.

We attribute the success of GIS to four key elements of the program model. The program is (1) *accessible*, with monthly meetings held on individual school grounds immediately upon dismissal, (2) *affordable*, due to the many women volunteers who enable the program to be offered at no cost to the girls or the school, (3) *flexible*, such that all monthly meetings are scheduled by

the school program coordinator to better meet the interests and schedules of the girls, and (4) *exciting*, due to the women scientists who engage the girls in age-appropriate, hands-on STEM activities each month. Without these elements, we do not believe GIS would attract such a large number of girls. In addition, we feel that this program model is readily transferable to intervention programs that target different age groups and topics and to mixed-gender groups.

References

Blue, J., and D. Gann. 2008. When do girls lose interest in math and science? *Science Scope* 32 (2): 44–47.

Catsambis, S. 1995. Gender, race, ethnicity, and science education in the middle grades. *Journal of Research in Science Teaching* 32: 243–57.

Farenga, S. J., and B. A. Joyce. 1999. Intentions of young students to enroll in science courses in the future: An examination of gender differences. *Science Education* 83: 55–75.

Gaston, B. 2001. STEPS: A tuition-free technology and science summer camp for girls. *Tech Directions* 60 (9): 20–23.

Jones, M., A. Howe, and M. Rua. 2000. Gender differences in students' experiences, interests, and attitudes toward science and scientists. *Science Education* 84: 180–92.

McCormick, M. E., and J. S. Wolf. 1993. Intervention programs for gifted girls. *Roeper Review* 16 (2).

Sax, L. 2006. *Why gender matters: What parents and teachers need to know about the emerging science of sex differences*. New York: Random House.

Stake, J. E., and K. R. Mares. 2001. Science enrichment programs for gifted high school girls and boys: Predictors of program impact on science confidence and motivation. *Journal of Research in Science Teaching* 38 (10): 1065–88.

CHAPTER 27

The Invention Factory
Student Inventions Aid Individuals With Disabilities

By Thomas W. Speitel, Neil G. Scott, and Sandy D. Gabrielli ///

The Invention Factory is a nontraditional youth-based, after-school program in Honolulu that teaches information technology and mechanics to teenagers through interactive, hands-on projects that improve human computer interaction for individuals with disabilities. The content area students study is electronics with embedded microcomputers, and the targeted population is students in grades 8 through 12. All teacher and student participation is voluntary.

One objective of the program is to stimulate interest in science and engineering careers among students currently underrepresented in those fields: women, native Hawaiians, students with disabilities, and at-risk students. Another objective is for students to learn enough electronics, mechanics, mathematics, and computer programming to conduct needs analysis, design, fabrication, and evaluation of devices that meet the needs of people who are disabled. The program intends to demonstrate that students who create technology-based solutions that impact people have increased motivation to pursue careers in engineering and science. The results are not in yet because the students are halfway through their three-year program.

An Idea for a Program

The idea for the Invention Factory program began with a single two-hour career-week workshop for high school girls. The organizers asked Sandy Gabrielli, a female engineer, to share her experiences in assistive technology (AT) engineering with the girls. Instead of just talking about engineering, she provided a short engineering experience by teaching students some elementary electrical theory and hands-on skills so that they could properly build a large switch and modify electronic toys to make them usable by children with motor disabilities.

The students, none of whom had used hand tools or soldering irons before, glowed with satisfaction when they successfully tested their switch and toy. This practical activity provided a perfect introduction for a short discussion about the engineering knowledge involved in the creation of more advanced human interfaces to information technologies. At the end of the workshop several students said they wanted to become engineers and all of them wanted to sign up for another workshop.

Later, student and teacher interviews revealed several different driving forces at work: The students were fascinated by the technologies, learned new information, mastered new skills, related to the intended user and recognized the need, understood how the proposed solution matched the need, and experienced a great sense of achievement when their small project was successful. These activities provided a way to get young people interested in science, technology, engineering, and mathematics (STEM). Word of this workshop spread quickly among teachers and within a few weeks there were requests from schools all around the island of Oahu for similar workshops to be held in schools. So from a small start, the Invention Factory grew.

Community Partners

This program consists of quite a few collaborating partners. Invention Factory staff are University of Hawaii employees from the colleges of education and engineering, including faculty, instructors, technicians, graduate students, and undergraduate students. A dozen undergraduates volunteer their time to help with the project. The Native Hawaiian Science and Engineering Mentoring program has provided the project with workshop instructional aides. Participating middle and high school teachers collaborate, providing school and curricular connections, supervision, and space.

FIGURE 27.1.

Examples of modules

Module 1: Introduction to Electronics
Students modify a toy and make a remote switch for children with motor impairments.

Module 2: Attention-Getting Devices
Students develop alternative alerts (attention-getting circuits) for people with sensory deficits. Students work with light (hearing impairments), sound (visual impairments), and vibration (hearing and visual impairments).

Module 3: Magnetism
Students build a Morse code communication device.

Module 4: Electrical Machines
Students invent a moving display to entertain and amaze.

Module 5: Audio
Students build an audio amplification system and voice-operated relay switch for people with speech disorders.

Module 6: Accessible User Interaction
Students develop a single switch scanner for operating a toy or a communication device for persons with limited movement.

Module 7: Introduction to Microprocessors
Students develop light and temperature sensor systems.

Module 8: Electronic Dice
Students build electronic dice to be used as a recreation tool for elderly individuals.

Module 9: Remote Control Applications
Students work with infrared and radio frequency to develop ways to control fans, lamps, televisions, and so on.

Module 10: Digital Recorder Applications
Students develop digital recorders to be used as simple sound- or speech-based games and communication devices.

Women in Technology, a nonprofit organization that encourages girls to enter the STEM fields, facilitates recruitment of participating schools. Gender equity objectives are addressed by attempting to have an equal number of boys and girls in each workshop, and mentoring opportunities are ensured for girls who participate in the program. Currently 34% of workshop students are girls and 19% of workshop students are native Hawaiians.

Community collaborations and projects for special-needs clients are initiated through staff contact with individual teachers, AT utilization specialists, and service providers. Outreach and publicity activities include presentations to teacher groups and project exhibits and workshops at local conferences for disability-related organizations. The program has grown in the community through word of mouth and successful experiences by end users, which generate additional requests.

Students have contributed over 100 modified toys and switches to the community in the first year of the project. The AT Resource Center of Hawaii maintains a lending library of toys that students have modified and provides students tours of their AT lab. Students have modified toys and donated them to Shriners Hospital and participated in recreational therapy with the hospital clients. Hawaii Department of Education special-education teachers, speech teachers, and teachers of the visually impaired have presented lectures about disability at workshop sessions and made specific device requests. Students then built those projects (predominantly toy modifications in early workshops) and gave them to teachers to use with students with special needs. Kapiolani Children's Hospital's Speech and Hearing Clinic provided students with project ideas that are being implemented in student design exer-

cises. The Hawaii Department of Health's Early Intervention Program has requested specific devices. In some schools, students are given design projects that involve students enrolled in the special-education classes at their own school. These projects are carefully defined to include the client student as an equal partner in the design and invention process.

Other community collaborators include the Engineering Information Foundation, the University of Hawaii's Institute of Electrical and Electronics Engineers student branch, the Special Parents Information Network, and parents, including home-schooling parents.

Promoting Inquiry

Invention Factory activities address numerous National Science Education Standards. As a result of their participation, students develop abilities in technological design and understandings about science and technology (NRC 1996, p. 190). Creativity, imagination, and a good knowledge base are all required for this work, and students respond positively to the concrete, practical outcome orientation of design problems. Students also see science in a social perspective, developing understanding of personal and community health and environmental quality (NRC 1996, p. 193).

Invention Factory personnel work closely with AT professionals to develop curricula that inform students as well as provide real-world applications and working public-service projects for student participation. The program curricula consist of individual modules that cover a basic concept in electronics; activities, exercises, demonstrations, and projects to support that concept; and an activity that directly relates to AT.

Instructors travel to schools with raw materials organized as kit sets designed and produced by Invention Factory staff. These sets and accompanying activity booklets help the instructors guide students through the modules (see Figure 27.1, p. 190, for examples of modules). Experience has shown that traditional lecture/lab format does not work with after-school program youth.

After some trial and error, the modules have evolved into sequences of short periods of hands-on activities, each preceded and followed by a short period of explanation or discussion. A typical class period may contain about six activities and demonstrations that teach basic principles

---

FIGURE 27.2.

Knowledge and skills self-inventory

Scoring

0 = Don't know or can't do it

1 = Know how to or have done it

2 = Can teach it to another student or teacher

I. Ability to recognize the following

Passive components

[ ] conductors   [ ] protoboards

[ ] insulators   [ ] connectors

[ ] lamps   [ ] resistors

[ ] batteries   [ ] capacitors

[ ] switches   [ ] inductors

Electromechanical components

[ ] motors [ ] relays [ ] microphones

[ ] speakers

Sensors

[ ] light-dependent resistors

[ ] optointerrupters

[ ] temperature-integrated circuits

[ ] pressure

[ ] infrared detectors

Active devices

[ ] diodes

[ ] transistors

[ ] light-emitting diodes (LEDs)

[ ] optoisolators

[ ] integrated circuits

II. Can explain the function of

Passive components

[ ] conductors   [ ] protoboards

[ ] insulators   [ ] connectors

[ ] lamps   [ ] resistors

[ ] batteries   [ ] capacitors

[ ] switches   [ ] inductors

Electromechanical components

[ ] motors   [ ] relays

[ ] microphones   [ ] speakers

Sensors

[ ] light-dependent resistors

[ ] optointerrupters

[ ] temperature-integrated circuits

[ ] pressure

[ ] infrared detectors

Active devices

[ ] diodes   [ ] transistors

[ ] LEDs   [ ] optoisolators

[ ] integrated circuits

---

that collectively lead to a functioning electronic device. Many of the activities are designed to evoke the question, "How could a disabled person use this, or use a variation of this?" Toy modifications are introduced during the first module as a lead in to teaching practical skills and applying

what students have just learned.

The first module consists of an introduction to electricity and simple circuits during which students construct a circuit consisting of a battery, lamp, and switch. After students complete wiring, we ask them, "How could you modify

---

FIGURE 27.2. (continued)

III. Skills

Using hand tools

[ ] sidecutter   [ ] solder sucker
[ ] hand punch   [ ] longnose pliers
[ ] wire stripper   [ ] metal press
[ ] wrench   [ ] hammer
[ ] hot-glue gun   [ ] screwdriver
[ ] hacksaw   [ ] needle and thread
[ ] hand drill   [ ] ruler
[ ] scalpel   [ ] electric drill
[ ] scribe   [ ] heat shrink gun
[ ] soldering iron   [ ] center punch
[ ] other _____

[ ] Reading an electrical circuit schematic

[ ] Using a digital multimeter to measure resistance, voltage, and current

Following a circuit diagram and assembly instructions to construct a hand-wired electrical device including

[ ] disability access switch
[ ] flashlight
[ ] modified electronic toy
[ ] other _____
[ ] Morse code key switch and buzzer

Following a circuit diagram and assembly instructions to assemble and test electrical circuits on a prototyping board

Audio circuits

[ ] microphone [ ] amplifier
[ ] speaker   [ ] tone generator

Digital circuits

[ ] logic [ ] timer [ ] counter [ ] other

Programming a microprocessor to perform a variety of tasks

[ ] flashing LEDs   [ ] controlling a motor
[ ] producing sounds   [ ] electronic die
[ ] other _____

IV. Invention

Describing an invention

[ ] name
[ ] microprocessor programming
[ ] general description
[ ] mechanical specification
[ ] electrical specification   [ ] sketches

Prototyping an invention

[ ] simulation   [ ] working model
[ ] trial test   [ ] user instructions
[ ] usability test

---

this to enable it to be used by a person with a disability?" After students have figured this out and modified the lamp circuit so that it can be controlled by a remote switch, they then assemble an accessible switch and test it with the lamp circuit. Then students select a toy, study how it works, decide which switch or switches need to be made remote, make drawings of what they propose doing, dismantle the toy, insert additional wires and sockets, reassemble the toy, and test it with the switch they made earlier. If the toy does not work, students backtrack to find the problem, fix it, and then reassemble the toy. All of these activities introduce and reinforce a wide range of hand, analytical, and invention skills. Both the toys and the switches are donated to disabled children.

Disability awareness lessons are included in each module. For example, in Module 1 (Figure 27.1), students are given additional reading about inclusion of students with disabilities in classrooms and have a class discussion about issues of accessibility.

Assessment and Self-Inventory

Much of the student assessment is informal. Because our primary goals are to develop positive attitudes toward STEM subjects and to stimulate inventiveness, any specific skills and knowledge students acquire are seen as a bonus. Later modules do include deliberate information gaps to be bridged by students, and at key points in any project's construction students must have devices checked by an instructor who will point out problems, opportunities, and related concepts. Most of the activities can be considered a success if the project is assembled correctly and neatly and the circuit or device works in the intended manner.

Students keep a record of their acquired skills, knowledge, and invention ideas and development in an engineering notebook. We are just starting to implement a concept-and-skill student self-assessment that can be corroborated by an instructor. The prototype is shown in Figure 27.2 (pp. 192–193). Important statements include invention of an accessible device, ability to recognize and give the function of electrical components, ability to solder, ability to design a circuit board, and ability to program a microprocessor. We are also developing a resume-writing computer program

FIGURE 27.3.

Radio-controlled car

Student invention to encourage a child with autism to participate in speech-therapy activities.

tied into the concept-and-skill student self-assessment that will help students communicate the wide array of skills, knowledge, and experiences that they have garnered in and out of the Invention Factory program. The authors believe that students will graduate from the program with the confidence and knowledge, as exemplified in the resume, to further pursue STEM careers.

Examples of Inventions

One example of the types of inventions developed through the Invention Factory program originated during a class presentation—a speech therapist described how a young boy with autism had refused to intentionally blow air. The child's parent had specifically requested in the individual education plan that the student be able to blow out his birthday candles; his teachers and the speech therapist had tried everything from bubbles to a silk scarf but nothing was working. Invention Factory students knew that radio-controlled cars are very motivating to many young boys, so they designed and built a working pinwheel interface made out of a small microprocessor, the guts of a computer mouse, a very cool off-the-shelf car, and a pinwheel. As the pinwheel is blown, the car drives forward or backward (Figure 27.3, p. 194).

Another group of students designed an "Incredibles Mobile" after hearing a class presentation from a family member of a five-year-old boy with autism. Students were told that the child loved things that are heavy, are circular, and have repetitive motion. The child needed functional-skills development in understanding cause and effect and how to turn knobs. Students designed and created a mobile out of a motorized musical baby-crib mobile, accessible switches, and action figures from the movie *The Incredibles*.

Impacts of the Invention Factory

The Invention Factory shows students that they can invent and produce practical solutions to real problems and makes them aware of entrepreneurial opportunities. By focusing on inventing solutions to problems facing disabled and aging people, students fulfill their community-service responsibilities and increase their awareness of how technology can be used to help people. Techniques developed to assist individuals with disabilities naturally fit the needs of elderly people with mental or physical limitations brought on by aging. Some students have made a trip to a nursing facility and met with elderly people. The project is just reaching the stage of increasing these activities. Applications for elderly people include turning on appliances instead of a toy, answering the phone, and electronic dice for playing board games. Students are taught the processes involved in taking an idea through to the product stage. They are shown the difference between a handmade prototype and the same thing assembled on a printed circuit. Students have the experience during the summer of designing and making a microprocessor unit on a printed circuit. They also begin to understand the value of intellectual property and are taught about the patent system and what it does to protect inventors.

The Invention Factory has five levels of impact: (1) students attending the Invention Factory; (2) the products they generate; (3) the disabled and elderly populations affected by students; (4) community awareness (parents, regular school teachers, and the general public) of the need for STEM education and the needs of people with disabilities by press coverage and word of mouth about the impact of student outreach; and (5) the reproducibility and extension of the program to other communities. One goal for next year is to increase the number of opportunities for students to meet directly with the consumers of AT.

Creating a device that empowers a disabled child to do something they have never been able to do before, or enabling an elderly person to regain the ability to do things they thought they would never be able to do again, is an extremely moving, exciting, and motivating experience. There is a natural feeling of wanting to come up with an even better solution that strengthens the incentive to learn how to do so. Seemingly irrelevant physics or mathematics activities suddenly become relevant to meeting real needs. Once the link to real people with real needs is established, the ability and skills to actually make a working device become just as important as knowing how to design the software and electronic circuits.

Acknowledgments

This project was supported in part by the National Science Foundation grant titled "Invention Factory." The authors would like to thank the Twizzlers student team for their contribution of the pinwheel controller documentation.

Reference

National Research Council (NRC). 1996. *National science education standards.* Washington, DC: National Academies Press.

CHAPTER 28

Engineering-A-Future for Tomorrow's Young Women

By Susan Gore ///

Engineering-A-Future (EAF) is an outreach program for middle school age girls. Through a collaborative effort of the Colleges of Engineering and Education at Tennessee Tech University (TTU), the participants experience hands-on activities to foster an interest in career options that are still considered nontraditional for females among elementary and middle school girls. The original format of EAF places participants into groups of 8–10 members that rotate through four different 45-minute hands-on activities located at four different sites.

The Program

As Craven and others assert, "It is well known that women are underrepresented in the engineering profession" (Craven, Pardue, and Ramsey-Idem 2005). One goal of EAF is to provide a strong role model for these young women in the field of engineering. A team-building activity, Reverse Engineering, was conducted at the beginning of the day. Each team was given a box of "parts" that had once been items such as a dot matrix printer, a toaster, a typewriter, a computer, a microwave, and so on. Many school districts or industries have outdated printers and computers that they are willing to donate for this purpose. Old electric pencil sharpeners, can openers, blenders, or other small appliances could also be used. The teams were given 10 minutes to determine what these parts had been originally. Step II of this activity involved taking these parts to design their own invention and explain how it would be used. See the Reverse Engineering activity sheet.

The Activities

There are too many activities to individually highlight, but while interviewing participants, one activity seemed to be a real hit—Edible Villages (see Edible Villages activity sheet). The

EDIBLE VILLAGES

What happens to buildings during an earthquake?
Record answers on your own paper.

Village I

1. From the supplies on your table, build identical
   one-story buildings on each of the different "soil"
   types (Jello-O and Play-Doh).
2. Before shaking each tray, predict what your group
   thinks will happen to the building on it. Record your
   predictions.
3. Shake the trays gently and record what happens.

Village II

1. From the supplies on your table, build identical two-
   story buildings on each of the different "soil" trays.
2. Before shaking each tray, predict what your group
   thinks will happen to the building on it. Record your
   predictions.
3. Shake the trays gently and record what happens.

Village III

1. From the supplies on your table, build identical
   three-story buildings on each of the different "soil"
   trays.
2. Before shaking each tray, predict what your group
   thinks will happen to the building on it. Record your
   predictions.
3. Shake the trays gently and record what happens.

All Villages

Visit each village to survey the earthquake results.
Then write your conclusions based on the building
structures and the soil types. What can you infer about
building foundations? Do you think the number of
stories is important to the safety of the building? Why
or why not?

purpose of this activity was to find out
what happens to buildings during an
earthquake—something our students
found particularly interesting because
earthquakes have occurred and do
occur in Tennessee. Sally Pardue, an
associate professor of engineering at
TTU, developed this activity based
on a presentation by Neda Fabris at
the American Society of Engineering
Educators Annual Conference and
Exposition (Fabris 2001).

Materials

- *Building materials*—stick-shaped
  pretzels, marshmallows, caramels,
  gum drops, and large spaghetti (use
  to spike the soft materials together).
  Amounts will vary depending on
  the creativity of individual classes.

- *Soil materials*—gelatin and play
  dough placed in disposable trays

The "soil" foundations need to be
made before the activity. The Jell-O
foundation should be made with
double the gelatin (as for Jigglers).
The challenge is to make a foundation
about 5 cm thick that is still pliable
enough so that it does not break when
building bases are inserted (see Edible
Villages activity sheet for details). This
foundation can be made several days
before if kept in the refrigerator. However, it can only be used for one class
period because holes in the surface
will weaken the foundation. The play
dough can be made from any standard
recipe of flour, salt, cooking oil, water,

and cream of tartar. These ingredients are safe for students to handle. Again, it can be made ahead of time if kept covered to prevent it from drying. It can also be used throughout the day. You will have to slightly press out the holes after each class use.

Safety Guidelines

Even though all the materials used are edible, students should be cautioned not to eat the building or foundation materials.

Management Tip

Cover tables with newspaper. As with any activity, this one will go more smoothly if everything is prepared ahead of time and placed in containers for each group. We have divided this activity into three groups. Depending on class size, group number may need modification. For example, each village could be split into two villages by building on only one soil type, making six groups instead of three.

Wrap-Up

Let students discuss their final results as a class.

Extending the Activity

Ask students how engineers and architects work together to make safer buildings. They can learn more online at the U.S. Geological Survey Earthquake Hazards Program (see Resources).

Parent Program

In conjunction with EAF, a program is conducted for parents or guardians of participants, as well as interested adults. Each adult attending the workshop receives a copy of the National Science Foundation's "New Formulas for America's Workforce: Girls in Science and Engineering" (see Resources).

REVERSE ENGINEERING

Step 1 (10 minutes)
Directions:
Your team has been given a box of parts that were once a useful, working item. Examine the contents of the box and complete the following sentence.

1. We think the item was a _____
because _____
_____
_____

Step 2 (15–20 minutes)
Directions:
Now your team must take these same parts and create a new invention. Complete the sentences below and be ready to share your new invention with the class.

1. Our new invention is called _____

2. It is used to _____
_____
_____

Topics discussed include

- preparing for success in college,

- funding a college education,

- estimating the costs of college education,

- emphasizing the importance of communication skills,

- emphasizing the need for precollege math and science courses,

- obtaining college credit prior to beginning college, and

- illustrating examples of various engineering curriculum (Craven, Pardue, and Ramsey-Idem 2005)

The Results

With the success seen thus far, the EAF organizers are planning for the future. The program will expand into a day camp that would coincide with a regularly scheduled school break. This will allow participants to attend for several hours a day, for four to five days. If this is successful, the program will be expanded into a residence camp during the summer, providing attendees housing on the TTU campus for several days. This would provide for more social interaction and also allow the organizers to expand the length, depth, and type of activities that can be implemented. Significant support will be necessary before this level can be attained, and as such, it is still in the somewhat distant future.

Acknowledgments

A core group of women faculty members and students from the College of Engineering and the College of Education at Tennessee Tech University, the Cumberland Valley Girl Scouts Council, members of the Society of Women Engineers, women engineer employees of Fleetguard, Inc., and members of the American Association of University Women collaborated to make EAF a reality.

References

Craven, K. K., S. Pardue, and K. Ramsey-Idem. 2005. Engineering a future at Tennessee Tech University. Paper presented at the 2005 American Society for Engineering Education Annual Conference and Exposition. Portland, OR.

Fabris, N. 2001. Earthquakes, materials and an edible village: An educational experiment for high school students. Paper presented at the 2001 American Society for Engineering Education Annual Conference and Exposition. Albuquerque, NM.

Internet Resources

National Science Foundation's New formulas for America's workforce: Girls in science and engineering
http://www.nsf.gov/pubs/2003/nsf03207

U.S. Geological Survey Earthquake Hazards Program
http://earthquake.usgs.gov/learning/kids.php

CHAPTER 29

A Model of the INEEL Science and Engineering Expo for Middle Schools

By Elda Zounar ///

Informal education can augment science classroom instruction and draw students to careers in science. The science and engineering expo, created by the Idaho National Engineering and Environmental Laboratory (INEEL), is a high-impact, inquiry-based, informal-education experience that you can replicate. It engages a broad range of students in learning and exploring science, mathematics, engineering, and technology. The expo is aimed toward youth in grades 5–8, teachers, and parents, and it is also of interest to community leaders and learners of all ages.

Your science and engineering expo project can be a large community-based project, a small schoolwide project, or a classroom project. The size and complexity of the expo is adaptable for middle school leaders. This model addresses National Science Education Program Standard D in that it gives students access to many community resources in both the short- and long-term (NRC 1996).

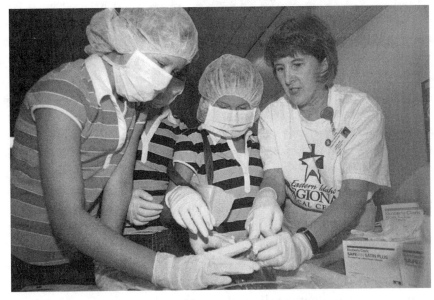

Wearing surgical gloves and masks, young surgeons at a medical exhibit investigate a fresh pig heart to learn how the human heart works.

Pilot

The science and engineering expo was conceived in 2000 at the INEEL and piloted in October 2001. From a community perspective, the expo was a homegrown, fun, safe, informal education event that was planned and offered by local leaders. The INEEL and the Idaho Science Teachers Association (ISTA) partnered to host the expo on a statewide teacher inservice day. The INEEL ran the expo. We wanted students to attend the expo on one side of the street while teachers attended the annual teachers'

This boy is fascinated by an industrial-size robot's manipulator and remote-handling characteristics.

meeting on the other side of the street. It was a venue option that was included in the program of activities for teachers at their annual meeting so that teachers could also attend the expo. We called it a Science Odyssey and Expo.

The expo brought together local community resources through exhibits, interactive demonstrations, simulations, experiments, presentations, activities, sponsorships, and volunteerism. We wanted diversity in subject matter to explore students' interests and variety in instructional techniques to draw attention to students' varied strengths (AAUW 1999). The expo showed how science and technology could be exported from a workplace such as a robotics laboratory, chemistry laboratory, or hospital surgical room, to the middle of a school gymnasium.

Exhibitors were outgoing adults who like to stimulate curiosity, inform, or teach. They were self-identified technical contacts from local organizations who served as resources for teachers throughout the school year. The expo promoted equal access to the information by being open and free to the public.

A Science and Engineering Expo Model for Middle Schools

This model for middle schools is academic-focused and designed around lessons learned since the pilot. Our lessons tell us that the best place to hold a large community-based expo is a conference center. In lieu of such a facility, a place associated with learning such as a school, university campus, or discovery center can be used. The proper time to hold an expo is during the school year so that teachers can give students credit for attending and so that students can see relationships between what they are learning in class and what they experience at the expo from practitioners in different fields. The effectiveness of the expo is maximized if it is linked directly to curricula and standards.

The objective of your expo program should address teacher and school needs while reaching out to the interests of students. Including your middle school students in the planning, implementing, and evaluating of the expo will add an important teaching and learning dimension to it. Here are 12 steps to expo success:

1. Find an industry or university partner. Corporations and science/engineering university departments are very education-friendly. Select one from the local area, make telephone or personal visit contact, and explain the intent of the expo. Their payback is the goodwill and name recognition they receive from the promotions and publicity. The partner should help with the strategic planning of the expo, help to defray costs, share the workload, and brand the expo with a logo.

2. Design one or more planning tools that will help you stay organized. Identify teaching

techniques and curricular objectives that could benefit from an informal education program. Link the objectives to your state or the National Science Education Standards. Figure 29.1 is an example of a simple standards-chain and how INEEL expo delivery activities have met standards (NRC 1996).

3. Explore local municipal, industry, university, and professional society opportunities that could help you meet the objectives and address the standards.

4. Recruit representatives from these agencies to partner with you to deliver science education. Register each partner and identify any equipment or other needs.

5. Establish a method of delivery for each science education partner. Large-group presentations, small-group demonstrations, and exhibits are three

FIGURE 29.1.

National Science Education Standards for grades 5–8

Program standards	B: The program of study in science for all students should be developmentally appropriate, interesting, and relevant to students' lives; emphasize student understanding through inquiry; and be connected with other school subjects.
Teaching standards	A: Teachers of science plan an inquiry-based science program for their students.
Content standards	A: Science as inquiry B: Physical science C: Life science D: Earth and space science E: Science and technology F: Science in personal and social perspectives G: History and nature of science
INEEL expo delivery activities	A: Presentation of scientific approach to the discovery of atoms and quarks. B: Chemistry magic show, bed of nails, and energy-generating bicycle. C: Medical exhibits, exhibits harnessing nature's diversity. D: Rocket launch; exhibits on geology, volcanoes, water, and soils. E: Swarming robots, cryogenics, and experiments. F: Mobile zoo, smoker pig lungs, and municipal water and sewer system models. G: Exhibitors are role models and practicing scientists.

potential methods of delivery (see Figure 29.2, p. 204). To ensure that presentations, demonstrations, and exhibits are top-quality and meet identified curricular needs, judge them for (a) technical merit, (b) visual impact, and (c) presentation impact, all in the spirit of delivering fun science for everyone.

6. Identify other activities, such as mathematics competitions, science bowls, featured speakers, and broadcasts, that can contribute to the expo.

FIGURE 29.2.

Methods of delivery

- **Large-group presentations.** These presentations are featured and scheduled. INEEL expo large-group presentations have included cryogenics, discovery of atoms and quarks, and the international space station.

- **Small-group demonstrations.** These demonstrations may or may not be featured and scheduled. They are timed and delivered repeatedly. INEEL expo small-group demonstrations have included chemistry magic shows, glassblowing, robotics, and canine officer demonstrations.

- **Exhibits.** Exhibit booths are subject-specific places of continuous teaching and learning activity. INEEL expo exhibits have included turkey breast biopsy, bed of nails, water and soil types, plants and animals of the desert, energy conservation, and natural disasters.

---

7. Establish a daily and hourly schedule for the delivery of the science program.

8. Identify budgetary needs and potential sponsors who can give you funds, materials, or services.

9. Identify personnel needs and recruit volunteers from citizens' groups, parents, and older students.

10. Perform a safety walk-through and address all potential hazards prior to opening the expo doors. Design exhibits around emergency staff so they are on-site if needed.

11. Communicate clearly and regularly with your partners, parents, community, and the media.

12. Evaluate. Design the evaluation so results tell you how effective your expo was and whether or not it should be repeated in future years. See Figure 29.3 for suggestions.

Classroom Expo

Even though the scope of an expo can be reduced to the size of a classroom project, the effectiveness of it does not need to be reduced. From the expo model, identify and apply selected elements. For a classroom expo, section the space into stations where students will be able to learn from the subject-matter experts in a variety of related subjects simultaneously. To deliver an expo that focuses on the National Science Education Content Standard F: Science in Personal and Social Perspectives (NRC 1996), for example, identify a few thematic threads such as animal and human populations, environmental impacts, and community health. Contact your local experts by e-mail, telephone, or a personal visit, and ask them to generate scenarios representative of your geographic area that illustrate the academic content of the standard and also explain the job of the expert. A sample set of delivery activities follows:

- Ask a zoo docent to bring different types of animals, birds, or reptiles whose unique characteristics highlight their survival abilities in their respective habitats.

- Invite an archaeologist to bring artifacts of the same or similar creatures as those from the zoo to show evidence of human, natural, or hazard-imposed environmental and cultural impacts on them over time.

- Have a master gardener bring a variety of living and mounted insects to explain their helpful and harmful effects on animal and human communities.

- Call on the water conservation district to provide buckets of different soil types and, through "hands-in" exploration, explain soil amendments, soil erosion, and why different types of environments are prone to wind and water erosion.

- Ask a geologist to bring fish tank models and show by injections of colored water how geologic features control rate and direction of water flow beneath the ground.

- Request the municipal water department that provide a functioning tabletop model of the city water system and show how water travels safely through residential infrastructure.

- Ask the municipal sewer department to provide a functioning tabletop model of a wastewater treatment plant and show how technology is used to process wastewater to the benefit of society.

FIGURE 29.3.

Evaluation examples

- Give "Passports to Science Knowledge" to students, who must present them and get them stamped at every activity or exhibit. Collect the passports at the end of the expo. A content analysis of the stamped passports will help you determine attendance at the different stations. Ideally, every student should visit every station.

- Quiz all students on the content of the expo activities since all activities should augment formal instruction. Responses to the test questions should indicate the extent to which students learned, at some level.

- Record focused observations of student behavior prior to, during, and after the expo. Supplement observations with photographs or videotape.

- Interview a sample of students for their perceptions of the expo.

- Survey all students on their academic and career interests based on what they experienced through expo activities. This information will be useful for future planning.

Conclusion

The pilot expo in 2000 turned out to be a teacher-pleasing, kid-winning event. More than 700 K–12 teachers and 2,500 students, parents, and citizens of the region attended. Nine hundred and one students completed evaluations of the expo. Twenty-one percent said they liked "everything" the best. Student interest was exhibited most highly in robotics technology, technology transfer, safety, energy, environmental education, and chemistry.

At the INEEL, we are cognizant of national economic and political pressures to encourage students into the science career pipeline. The INEEL science and engineering expo will continue to be a deliberate effort from an industry-partner perspective to introduce upper elementary, middle, and junior high school students to a wide and diverse range of potential careers in science, mathematics, engineering, and technology. Using extensive local resources in science, engineering, and information management, the expo lures and hooks kids on learning. Metaphorically speaking, it is a means to transfer enthusiasm for learning from one generation to the next, thereby handing off a world of knowledge. Our hoped-for return is an informed and interested young citizenry.

Acknowledgment

Special thanks to Laura Eder, past president of the Idaho Science Teachers Association.

References

American Association of University Women (AAUW). 1999. *Gender gaps: Where schools still fail our children*. New York: Marlowe & Company.

National Research Council (NRC). 1996. *National science education standards*. Washington, DC: National Academies Press.

CHAPTER 30

Engineering in the Classroom

By Kathleen Matthew and Stacy Wilson ///

When asked, "What does an engineer do?" students usually answer, "It's the guy who drives the train." Statistically speaking, students do have the "guy" part correct, but engineers do more than drive the train. What most students don't know is that engineers are female or male scientists who apply physical science to their daily lives by designing, building, and maintaining trains, as well as generating ideas to improve their efficiency.

In an effort to educate students about engineering, teachers in the Bowling Green, Kentucky, area challenged their students to become engineers. Students were given the tools they needed to design, create, and race a vehicle constructed from plastic building blocks that would move down a predetermined course in the shortest amount of time. With a robotics competition as the culminating experience, upper elementary and middle school students were motivated to become engineers.

Training the Engineers

The first stage of the competition was to expose students to the profession of engineering. Students brainstormed questions they might like to ask a professional in this field. An engineer from a local manufacturing plant was invited to speak to students about engineering as a profession. Because the Corvette plant is in our area, we contacted the facility and asked if one of their engineers could come to our class to speak about engineering. We specifically asked for a female engineer in order to counteract gender stereotyping.

Teachers could contact the local sections of professional engineering organizations such as IEEE (Institute for Electrical and Electronics Engineers), ASME (American Society of Mechanical Engineers), ASCE (American Society of Civil Engineers), and SWE (Society of Women Engineers) for names of engineers in a given area. The local group of professional engineers would also be able to provide a list of potential speakers for teachers.

Students asked the engineer from Corvette many questions about what an engineer does. After she left, students investigated three different types of engineering (electrical, civil, and mechanical), using the internet as a resource.

Students observed a high school robot competition at Western Kentucky University's (WKU) engineering lab. Although the high school competition demanded a higher level of programming and construction, students came away ready to build their own robots.

In areas that lack a local high school robotics competition, teachers can put their own competition together. Some of our teachers reportedly had their students compete in the hallways, libraries, and cafeterias. Their students programmed robots to race down the hallways in the quickest time, make a square around a library table, and avoid obstacles like trash cans and milk cartons in the cafeteria.

Starting the Engines

Because we were working on programming and building robots, we focused on electrical engineering. In an activity to introduce the topic, students experimented with series and parallel circuits (see directions in Figure 1). Alan Bartholomew's book, *Electricity, Gadgets, and Gizmos: Battery-powered, Buildable Gadgets That Go!* (1988) provided additional activities for both individuals and teams.

FIGURE 30.1.

Experimenting with circuits

Materials
- One D cell battery (1.5 v)
- Two 6 in. lengths of wire (24 gauge, single strand)
- One flashlight bulb (1.5 v)

Directions
1. Put students in groups of two or three.
2. Give each group the above materials.
3. Ask students to light the bulb.
4. Give students time to experiment.
5. Provide a diagram or picture for those who need help (but only after they have experimented).
6. Next, give the groups an additional bulb and two wires. Ask students to produce a parallel circuit.

(For a list of websites with introductory activities, see Internet Resources.)

Our students worked in groups of four, but other group sizes would work well also. The investigations conducted by students helped them learn how electricity flows through series and parallel circuits. When the electrical system was set up and connected correctly, the lightbulb lit up. Immediately, students knew they had connected the wires properly.

For ease of materials management, resealable baggies containing a laminated materials list, a battery, bulbs, and wires for the investigation were prepared ahead of time. This made it easy to distribute the materials to each group and to clean up. Teamwork was needed in each group to light the lightbulb. Each team should work at a table or central desk so all students have access to the materials. Students produced both series and parallel circuits. The materials for this experiment can be reused; however, the batteries would eventually need to be replaced.

Later, it would be necessary to do the same for the robot. Students have to set up and connect the system correctly to get the robot to move.

Blowing the Whistle

Students were now ready to plan, create, and program a robot for an areawide competition using the robot kit. The kit comes with a large number of building components in various shapes and sizes, motors, touch sensors, light sensors, a small RCX processor (called *the brick*), and the necessary software.

The software has an easy-to-use tutorial, which teaches students how to program the robot. The program is written as a flowchart so that students are basically learning the fundamentals of computer programming while programming their robot. The brick is simply the brains of the robot and has locations to attach input signals (sensors) and output signals (motors). The computer program is easily downloaded to the brick, which will then control the actions of the robot.

Professional engineers mentored student groups throughout the planning and building phase, university faculty assisted students with the computer programming, and classroom teachers guided the entire process. This three-way partnership created a community of leaders focused on team building, problem solving, and student learning. It was a win-win situation.

Engineering mentors provided expertise and visibility for their profession. In most cases, the engineers were involved through networking. One engineer contacted by a teacher provided the names of other engineers. The mentors' names and affiliations were included in the competition's program, which provided a successful public relations bonus.

If professional engineers are not available in your area, teachers can mentor/coach a team alone. Parents, Boy Scout leaders, and Girl Scout leaders can be coaches as well.

Competing

Engineering students and faculty at WKU designed and built the robotic competing surface that we used, but elsewhere teachers constructed a surface on a tabletop or floor using black or white electrical tape as a line for the robots to follow. Beginning two weeks before the official competition, student teams were able to view, take pictures, make measurements, and ask questions pertaining to the surface. Each robotics team was allowed to test their robot two times on the official course during the week before the competition and to make programming adjustments based on their test runs. The number of students per team usually ranged from 2 to 10 and was left to the discretion of the coach.

Each team had eight weeks to build and program their robots before they competed. Teachers reportedly used both in- and out-of-class time for this project. In some cases, parents volunteered their time and even built a practice surface for students, which was modeled after the competition surface. Teachers not involved in an areawide competition may opt to compete within their school or their classroom as a culminating activity.

The competition was free to the public. On the day of the competition, spectators poured into the gymnasium: students, teachers, parents, principals, professors, and engineers, as well as members of the community. During the course of the competition, excitement and team spirit remained at a high pitch. Pride and determination registered on students' faces as the winners were announced; the teams bounced onto the floor to accept their trophies, signaling the end of a successful project.

Crossing the Finish Line

Parents, teachers, and mentors agree that the competition was an exciting success. A father of one participant was heard saying, "I did not know my daughter could build something like that." A teacher commented, "The kids learned so much." A team mentor announced, "This is great, we'll do it again next year and win." For these students, science is learned best when experienced; they now know that engineers design the train, as well as drive it.

Catching the Caboose

An NSF grant and a Kentucky Eisenhower grant covered the cost of the classroom kits for the first year. The cost for one kit is approximately $249. After the funding stopped, teachers obtained robotics kits for their classrooms in several different ways. Some teachers received the kits through their school, while others relied on their parent-teacher organizations to purchase them. A few teachers had parents or businesses donate a kit to their classroom. One student's parent organized a bake sale and car wash to raise the money for the kits. Raising the funds for the competition becomes a collaborative effort.

References

Bartholomew, A. 1988. *Electricity, gadgets, and gizmos: Battery-powered, buildable gadgets that go!* Buffalo, NY: Kids Can Press.

Glover, D. 2001. *Young discoverers: Batteries, bulbs, and wires.* New York: Kingfisher.

Van Cleave, J. 2006. *Energy for every kid.* Hoboken, NJ: John Wiley and Sons.

Internet Resources

American Society of Mechanical Engineers
www.asme.org
American Society of Civil Engineers
www.asce.org
Institute of Electrical and Electronics Engineers
www.ieee.org
National Society of Professional Engineers
www.nspe.org
Robotics kit internet resource pages
www.lego.com, *www.pasco.com*
Society of Women Engineers
www.swe.org

Index

*Page numbers printed in **boldface** type refer to figures or tables.*

Index

212

Index

Index